Reading Trout Water

2nd Edition

Dave Hughes

STACKPOLE
BOOKS

Published by
STACKPOLE BOOKS
5067 Ritter Road
Mechanicsburg, PA 17055
www.stackpolebooks.com

Printed in Singapore

10 9 8 7 6 5 4 3 2 1

Library of Congress Cataloging-in-Publication Data

Hughes, Dave, 1945–
    Reading trout water / Dave Hughes. — 2nd ed.
        p. cm.
    Rev. ed. of: Reading the water / Dave Hughes. 1988.
    Includes bibliographical references and index.
    ISBN-13: 978-0-8117-3644-2 (alk. paper)
    ISBN-10: 0-8117-3644-X (alk. paper)
    1. Trout fishing. 2. Fly fishing. 3. Trout fisheries. I. Hughes, Dave, 1945–
Reading the water. II. Title.
    SH687.H76 2011
    799.17'57—dc22
                            2010002028

Contents

Introduction

Ten percent of the fishermen catch ninety percent of the trout. It's an old saying, and it's probably wrong, but it's not too wrong. Twenty-five percent of fly fishers most likely catch seventy-five percent of all trout that get caught. Those who learn how to read water, locate trout, and present flies to those trout in a lifelike manner catch far more than those who fish the water at random. The reason is summed up in another simple old saying: *Ten percent of the water holds ninety percent of the trout*. The percentages are probably off on that one as well, but again, they're not off by a lot. It's likely that a quarter of the water holds three-quarters of the trout. You know what will happen to your catch if you learn to focus your fishing on that small portion of trout water that holds an outsize proportion of its trout.

My brother Gene and I once floated a wilderness river for several days with Montana guide Don Williams. I'd been having fair luck with wet flies, fishing them patiently on cross-stream swings, stepping and casting down the lengths of a few long runs, catching some satisfying brown trout and rainbows. We got behind schedule, and Don became anxious to cover some miles, but when we approached another run shaped perfectly for swinging the wet fly, I whined until he brought the drift boat to ground in the riffle at the head of the run. I hopped out and threw spray into the air in my eagerness to rush downstream to where the water smoothed out and looked best to me. I worked out line, began making my quartering casts, and was soon into the rhythm of it all, gazing up at swallows swirling around the cliffs while I cast, mended, and waited for a trout to rap my fly.

I fished two hundred feet of water before I felt a tap, then a strong tug, and a trout was on and in the air. Gene had waded in downstream from me and was fishing the same type of water the same way, on my advice. He was a beginner in those days and often did what I told him to, though he's learned better since. I asked him to come up and take my camera; I wanted to get pictures of this sixteen-inch trout. It was a brown, nicely shaped, and prettily spotted.

A minute later I released the trout, and Gene handed the camera back. Just then we heard a shout. We looked upstream and there stood Don in a foot of water ten feet from the boat, playing a trout that was bigger than

Reading water is what you do to predict the most likely locations of trout in moving water in order to concentrate your casts right *there*, so you can maximize your chance to catch one, or a bunch, or some satisfying number in between. When you've spotted a rising trout, or a pod of them, you've succeeded in locating trout. Your exercise in reading the water should cease at that point, and you should start off on the studious set of steps that leads to catching them: Do your best to ascertain which stage of what insect or other animal the trout might be eating, frisk your fly boxes to find the nearest match you own at the moment, and move into the best position from which to present that imitation so it arrives in front of the trout looking and moving like the natural they're already eating.

Looking for rises is a skill you should hone as you learn to read trout water. Always scan the water as you move along it. Watch it carefully for at least a short while before launching a first cast over any new bit of water. Peer closely, and if your eyes are no better than mine, don't hesitate to cheat with binoculars. A bold rise on a flat like the one in this photo might be easy to spot; most rises are much less visible. Watch with special care over water that is even slightly rough; a run or riffle can easily conceal delicate rises. The smallest set of concentric circles on the water can be set in motion by the largest trout sipping tiny insects with delicacy and delight. Small rises do not always denote small trout.

the one I'd just released. But it wasn't much bigger, and I'd just had some success of my own, so I turned my back on him and returned to sweeping that long, unfeatured run with a wet fly. It all looked good to me.

Spotting Trout

Sometimes a trout will expose itself whole to you, easily seen, starkly visible on its lie. You'll be able to watch it at work, see whether it's feeding or resting, and if it is feeding, tell whether it's intercepting nymphs along the bottom or lifting to feed on something in the mid-depths or subsurface. If you keep out of its sight—remember, if you can see it, it can likely see you—then you'll be able to observe it and plot against it. Suspend your activities in the direction of reading the water until you've dispensed with the spotted fish or it has dispensed with you. Focus on that trout and do whatever you need to do to catch it.

Trout will not often expose themselves to easy observation. Those anglers who are best at spotting them, and I am unfortunately not among them, are able to notice

parts of them in opposition to the water in which they live: the square shape of the tail against rounded bottom rocks, the white flash of an opening and closing mouth, a shadow beneath an invisible trout, or the vague outline of a fish, as in the photo to the left. Wear polarized sunglasses for most of your fishing and get into the habit of peering into trout water looking for fish, or parts of them. Those who are most able to spot trout do not have some sort of instinct that sets them apart from the rest of us; they're simply the folks who have the most experience trying to spot trout and have therefore built the biggest memory banks about what trout look like beneath the water. It's wise, on all types of water, to look for trout wherever it's remotely possible to spot them.

Some sorts of water, and some sizes of trout, are more readable than others. In New Zealand, where most trout streams are clear and most trout are large, the most common way to read water is to look for trout and find a fish before a cast is ever made. In small-stream fishing, where the trout typically are not large and the surrounding forest casts concealing shade over many likely lies, you'll rarely spot a trout before casting to it.

Gene didn't return to his fruitless fishing. He didn't know any better than to run up there to see what Don was doing. In no time, I heard him shouting. I was expecting to have a fish hit at just that moment, so I ignored him and concentrated on the swing of my fly. A few minutes later, Gene and Don were both shouting. No fish had hit my fly, so I gave up, reeled in, and jogged upstream to see what was happening.

It was pandemonium.

A brisk riffle plunged whitely into the head of the run right where Don had parked the boat. When Gene and I scooted off downstream to fish wet flies, Don watched us a while and then picked up his rod and lobbed a weighted Yuk Bug or Bitch Creek or some similar, heavy and ugly nymph into the frothed water at the foot of the riffle, where it eased out into the run. The current delivered the nymph into the slower water at the head of the run, where a brown trout sprang on it. The cast was a long one, perhaps fifteen feet.

When Gene joined Don and, on his advice, pinched a split shot to the leader above his wet fly and made a similar cast, it had a similar result. Don released his first trout, made another cast, and that's when I heard them both shouting at once.

I fiddled with the camera while they played trout. Don's went over two pounds, and Gene's was closer to three. When they'd both released them, I tucked the camera away and turned around to trot back toward my own end of the run, but I hadn't gone a dozen steps before Don began to shout again. He'd dropped that portly nymph into the swift current off the end of his rod, almost at his feet, while he stripped out enough line to make a real cast. A trout attacked and hooked itself. I gave up and stayed there to take pictures. To my misfortune, there was not room for three to fish in that small spot. The two of them took about a dozen fish in the next half hour. The largest was just over three pounds.

All of those trout held in a constricted bit of water over a trench in the bottom, marked by a slight darkening and flattening of the rumpled water on the surface of the riffle. This holding lie was no more than twenty feet long, nor wider than fifteen feet. Downstream from the trench, the trout population thinned out into the kind that would let a fellow fish two hundred feet to take a single sixteen-inch fish.

When Don had Gene and me back in the drift boat and was making miles again, I asked him what had prompted him to pick up his rod and try that spot.

"It looked like it might hold trout," he answered, which sounded like only part of an answer to me. Don fished just half a dozen times on that five-day float. Every time he did, he got into a pod of trout. I would have been glad to attribute his luck to day after day spent floating and fishing the same water, but he hadn't floated that particular river in several years. He simply knew how to read water and find trout, no matter where he was.

Certain characteristics about that small piece of water where he'd parked the boat hinted to Don that he would find trout right there. The broad riffle upstream would deliver a richness of aquatic insects, larvae and pupae and nymphs, tumbling into the head of the run downstream. The rumpled water at the head of the run would render trout invisible from above, safe from overhead predation. The slight darkness and smoothness of the surface indicated a trench in the bottom, where the water would be a bit deeper and slower than the water in the riffle all around it, giving trout shelter from a constant battle against pushy currents.

The place had all the signs of a prime lie. It was, to Don, clearly marked as the ten percent of the water that would hold ninety percent of the trout in that stretch of river, or at least the twenty-five percent of water that would contain seventy-five percent of them. I had galloped down to fish the remaining percentage and from it had taken my allotted share: the single brown that, in the end, represented about ten percent of what we caught in that riffle and run before we launched the boat and drifted off to see more of Montana.

If you learn to read water and recognize the small percentage of it that holds the highest percentage of trout, you might not suddenly begin catching ninety percent of the trout, dazzling all your friends. But you are certain to begin catching more, and often larger, trout.

Needs of Trout

Trout are like folks: They have a list of basic needs, they are most comfortable where those needs are met, and that's where you'll find them hanging out most of the time. *Reading water, simply defined, is learning to understand the anatomy of a trout stream so you can recognize the ways different parts of the stream meet the basic needs of trout.* You'll find trout where their needs are met. You'll find an absence of trout where their needs are not met. This is why twenty-five percent of the water holds seventy-five percent of the trout: Only a small proportion of any trout water meets the needs of the trout that live in that water.

Each stream is unique; each has its own anatomy. A mountain creek, a foothill stream, and a large, mature river all have separate prescriptions for filling the needs of trout. A placid spring creek is shaped far differently than a brisk freestone stream. But there are repetitive themes from stream to stream, no matter its size or character. Certain water types—riffles, runs, pools, flats, pocket water, bank water—occur in varying proportions within all the stream types, and trout always hold where they do in accordance with the way those types of water meet their needs.

A trout needs five things to ensure its survival: shelter from the constant force of the current, protection from predators, water in the right temperature range and containing sufficient oxygen, adequate food, and a bit of territory in which to live. They also need currents over the right-size clean gravel in order to spawn, but learning to recognize spawning beds will help you locate them only when they're busy reproducing, at which time you should refrain from fishing for them.

SHELTER FROM CURRENTS

Shelter from fast currents is the first and most basic need of trout in moving water. H. B. N. Hynes, in his definitive *Ecology of Running Waters*,

Reading Water in a Blink

In his book *Blink,* Malcolm Gladwell introduced the concept that the best decisions are often made in an instant, with a single glance at a situation. His idea has a surprising amount to do with reading trout water. An experienced angler looks at a bit of water and in a flash recognizes the most likely location not only of its trout, but the most trout and often the largest. It seems a bit of a mystery, and it might be one, even to the person who has read the water and performed that instant bit of intuition. But it's far from an accident. The ability to look at water and locate trout at a glance is related to experience fishing either that bit of water or, more likely, lots of bits of water that looked a lot like it.

Trout, in streams of all sizes and shapes, hold and therefore take your flies in water that meets their needs. Water that meets the needs of trout has certain aspects of current speed, depth, availability of shelter, plentiful food, obstructions to the current, and some etceteras. These can all be observed in the shape of the water. Your conscious brain might not be trained to sort them all out. Mine is not. But your unconscious brain, usually referred to as the *subconscious*, though according to Gladwell, that is not quite accurate, takes everything in and builds up a memory bank of what water looked like where you hooked trout in your past.

The longer your past, the more memories you've built up and the more quickly and accurately you'll read the water. So the book you're reading at present contains your assignment to go fishing more often and create more of those memories. Reading trout water, at its simplest, is a matter of spending lots of time on trout water, observing it, casting over it, letting your brain sort out what the water looked like where trout rose up out of it to inhale your dry fly or remained concealed along the bottom of it but intercepted your dead-drifted nymph.

These two photos of the Futaleufu River in Chile remind me of this concept. My wife, Masako, who hasn't fished as much as I have, nibbled with a dry fly at the shallows inside that riffle for about fifteen casts. After each one, I told her, "Farther out." She walked her casts outward and finally placed her fly near the deeper water at the edge of the riffle. "Now you're going to hook one," I said, and a trout came up and took her fly almost as punctuation at the end of my sentence.

"How did you know?" she asked.

"Because you finally put your fly where a trout would be." It was a lot more complicated than that. It was all based on what Gladwell said in *Blink*. But I didn't have a conscious thought about it. You have my permission to put this book aside and go fishing, though I hope you'll come back and finish reading it later.

noted that "fishes are easily fatigued so even streamlined species which are well adapted to fast-flowing water cannot swim rapidly for long periods. The maximum speed of swimming of Kamloops trout in fact falls from 440 to 89 cm./sec. after only three minutes of continuous swimming, and all fishes spend most of their time resting in shelter." If a trout on an aquatic treadmill begins to get worn down after just three minutes, it would not last long facing the full force of a brutal, or even brisk, current. It would have to constantly swim upstream in order to stay in one place and would quickly get tuckered along the way.

This does not mean that trout will be found only where the current is nil. We've all seen fish feeding in fast water, and we've all caught them from riffles or runs where it was difficult to brace ourselves against the force of the flow. In fact, it's the movement of water that delivers necessary food to fish. They need current, but they are most comfortable, and will usually hold, where that current is least.

Current is least wherever it's deflected. Even in the fastest flows, an obstruction will create a pocket of slow or almost still water. Many, though not all, of the tip-offs to good holding lies are visible breaks in currents that would otherwise be too fast to hold trout. Boulders, shelves, ledges, and even rooted plant beds are all examples of obstructions that deflect current and form holding lies for trout.

Trout lies in moving water are almost always related to some sort of obstruction to the current, behind which they can find some shelter from that same current. Not many lies will be as obvious as this one downstream from the base-rock shelf on Pennsylvania's Big Fishing Creek, so you'll usually have to hone a bit sharper eye to notice where the water is deflected or slowed by some obstruction in its flow. You should also notice that the water just upstream from the same shelf builds up and slows before dropping over it. A trout, or perhaps two or three of them, are just as likely to be found in that flat bit of water as they are in the current below it that Masako is fishing.

Not all factors that calm flowing waters and create holding lies are visible. The friction of current moving over the bottom slows the water in a layer from a few millimeters to a few inches thick. The rougher the bottom, from gravel through cobble to boulders, the more friction it produces and the more likely a trout will find a comfortable lie in the midst of water that seems too fast for it to survive.

Turbulence is another invisible factor. On the surface, where a boulder might protrude, it's easy to notice eddies and swirls where the current separates to go around the midstream rock and rejoins downstream from it. The same thing happens along the bottom, where the population of rocks is a lot more dense, but you can't see it happening down there, though you might be able to see the submerged boulder that causes it. Water flowing over a bouldered bottom gets into all kinds of conflicts with itself, one vigorous tendril of current canceling out the other. The result is a patchwork of quiet places along the bottom where trout find comfort, even in some of the fastest water.

The larger the rocks and boulders on the bottom, the more friction and turbulence they cause, and the more holding water trout will find. Fortunately for fish, and for fishermen, the laws of hydraulics deposit the finest material, sand and silt and pebbles, in the slowest water; the largest material, rocks and boulders, in tumbling water. The faster the water, the larger the material on the bottom and the greater the likelihood of lies where a trout, or a bunch of them, can avoid fighting the current.

Quiet places where trout can rest are found even in the fastest water. But if all other factors were equal, they would seek sections of stream where the water was quietest.

All other factors are rarely equal, however.

PROTECTION FROM PREDATORS

The second need of trout, protection from predators, is one of the primary factors that keep trout from basking in the luxury of the easiest water. If trout always sought the quietest water, they would also be exposed to the most predation.

Trout predators fall into three basic types: birds, four-footed animals, and anglers. Kingfishers are obvious along trout streams, but they take mostly smaller trout and perhaps deserve a few in trade for the beauty they bring to streams. Mergansers travel in damnable family flocks and are rapacious. On large rivers, they might be wonderful to watch, but when I see a fleet of twenty scouring one of my favorite small streams, I'm forced to remind myself that they are an elemental part of nature, they got there before I did, they've helped hone the trout into something I love, and although they seem to be eating the trout I'd like to catch, they belong there more than I do.

Herons can become a factor in trout populations on the smallest streams, though I think only when conditions get out of whack. They're also a part of nature, and when everything is in balance, their predation, though probably not slight, is not likely to be fatal to the abundance on which they depend for a living. Herons used to be a rare sight on one of my favorite local creeks, but after loggers stripped all the cover from its banks a few years ago, before regulators began requiring that buffer strips be left along streams, I would see at least one or two of these long-legged birds along the creek's course in every day's fishing. Logging silted in the spawning beds in the stream and greatly diminished recruitment of small trout to the population. But the sudden exposure to herons was likely the finishing shot; the creek became almost barren of trout.

Ospreys are the epitome of aerial predators of fish. Yet anyone who has had the fortune to watch one work—with its easy flight, plunging dive, detonating splash, and lumbering liftoff with the prey, perhaps overweight and difficult to grip, flapping in its talons—would never begrudge an osprey the few fish it takes. Sometimes the *many* fish it takes.

In recent years, I witnessed how the absence of overhead predation can make a big difference in the places trout hold. Patagonian Chile and Argentina had no native trout, and no natural large aerial predator—no osprey or fish-eating eagle—evolved in that environment, though kingfishers take their tithe of the trout that have been introduced. Once those trout get large enough to be safe from the kingfishers, however, they no longer have any worries about overhead predation, and they seem to have figured it all out in the 150 years or so they've been down there. Big trout hold in thin water right along grassy banks that are not undercut. They lie exposed in direct sunlight, in water just deep enough to cover their dorsal fins, in places where it ordinarily would be unusual to find them.

If you fish where trout are exposed to overhead predation, which most of us do most of the time, you need to know that trout will rarely hold carelessly on stations where they could easily be spotted and snatched from above. They desire some depth, to conceal them and limit their chances of being in reach of something smacking into the surface. They will hold up higher in currents that are rough on top, because it's more difficult for a bird's vision to penetrate rumpled water. They'll tuck themselves into the slightest shade, under any overhead cover, back beneath an undercut that keeps them out of sight of birds perched on branches, stalking the edges, or wheeling overhead. When you learn to look for the kinds of hides that the eyes of birds are unable to penetrate, you'll almost instantly begin to sort out holding lies from water without trout, especially where the water is less than a foot or two deep.

Otters and mink are the primary four-footed predators of trout. Otters are extremely effective. My wife and I watched a pair working an estuary once; they came up and crunched a small flounder after almost every dive. But that was an estuary and those were flounder. I doubt if enough otters or mink hunt the edges of flowing water to be a devastating factor in many natural trout populations, not today. Perhaps it's unfortunate; I'm delighted whenever I catch sight of an otter or mink looping along the shore.

I'm never sure how the presence of these swimming predators affects the distribution of trout. I suspect they make trout wary, but I don't think they cause trout to change their lies except in temporary ways, such as flight on sight. It's possible that the continual presence of a family of otters in one location would move trout out. I've never seen that happen, however, and otters seem to be more migratory on waters I fish: here today, somewhere upstream or down tomorrow. But my experience with them is fairly limited.

I have had trout feed, and even take my flies, when otters were near. Once when I slipped down to fish some bank water on the Deschutes River, I noticed an otter slide off a rock and dive away from my own predatory presence. I nymphed the water and soon hooked a nice trout. I'd about defeated it and was beginning to lead it toward shore to land it, when the trout suddenly gained massive strength, bolted directly away upstream against a strong Deschutes current with a hundred feet or more of my backing, and left me hung on a bottom rock. I never did see that otter again, but I've always felt it owed me an apology.

I fish a series of ponds that were installed like beads on a necklace high in the seepage headwaters of a tributary to the Deschutes River, out in wheat lands where the ponds are considered favorable because they stabilize flows over steelhead spawning beds much farther down the watershed. These ponds have no natural spawning, and because they lacked predators, they at once became incredible incubators for aquatic insects and still-water crustaceans. They've been stocked with trout that grow to good size almost instantly in such a rich environment.

Otters are explorers. It didn't take a family of them long to nose their way up the tributary, into and out of and all the way through to the highest of those new trout ponds. They knew precisely what to do about such a concentration of their favorite groceries. I've been asked by the rancher to please shoot them, but aside from the illegality of such brutalities, I've neglected to kill any because their presence makes me think the whole operation is tending a bit toward the direction nature would intend for it, though it will never be natural.

That leaves us: you and me. It's our job, if we are to take trout, to appear not to be predators to them. The first thing to avoid is any wading waves that would signal your presence to the trout. The second is any knocking together of rocks or thumping of boat bottoms and oars that would warn them you're coming after them. The third, and perhaps most important, is any flash of rod or line over their heads that might remind them about birds up there. Nothing will put trout down more quickly than the sight of something sailing overhead.

Protection from predators, for trout, can come from several sources. One of the best is water that is rough on top. Birds can't see through a broken surface to find fish. If they can't see them, they can't capture them. So a fish under a foot of choppy water is a lot less nervous creature than a fish quivering over the bottom of a foot-deep glassy flat. That is the main reason it's easy to approach trout in a riffle but difficult to move onto them in a spring creek glide.

Shade makes excellent protection, especially when the day is bright. Contrast between dark and light makes it hard for the eye of a predator to perceive anything in shadows, unless the predator has slipped into the shade with its prey. I once spotted a pod of trout working in eight inches of water at the edge of a Deschutes River flat. They were in the half circle of shadow cast by a bankside alder tree, grubbing the bottom for nymphs, and I was out in the sun. I never would have noticed them if one had not gotten overeager and broken the surface with its tail when it shot forward to grab a dislodged insect. That was its undoing. I moved into the shadows with it and caught it. In my youth, I was a more voracious predator than I am now, and I ate it.

Depth is the most obvious protection from predators. The deeper the water, the more wavering currents conceal the trout and the less light penetrates down to them. Darkness conceals trout. They feel the most security in deep pools and as a consequence are least wary there. But any water that is even slightly deeper than the water all around it attracts trout. A trench that is just two or three feet deep is the most likely place to find resting trout when it's located on a broad flat that is only a foot deep.

Camouflage serves as excellent protection for trout. It's a form of concealment they use constantly. Their anatomy aids this adaptation: Their bellies are light and blend with the sky when viewed from below; their backs are dark and spotted, blending perfectly against the bottom when viewed from above. I sat for lunch next to a favorite mountain creek one bright summer day. The sun slanted in, igniting a pool almost at my feet. I watched it for signs of trout and finally spotted a couple hanging in the cover of an undercut boulder. Before I stood up to fish for them, I scanned the tailout with my binoculars. It was water I'd already been watching for several minutes.

I saw the largest trout in the pool—which looked like a monster through the binoculars, though it was not—less than twenty feet from me. It hung in the slightest crease of current, just above the pebbled bottom, its speckled back and sides almost glaringly exposed, yet nearly invisible, blending perfectly with the stones beneath its belly. It did not move except to take an occasional insect, and it was almost invisible without binoculars except in the moments that it moved. The wavering of its body, when it held its station, was masked by the wavering of that tendril of wrinkled

Camouflage is a trout's best friend and sometimes lets a fish hang out in open water, where you'd think it could be easily seen. More often, trout hold along or just above the bottom, and their colors blend in so well that they seem mere modulations of the current as it flows along. This brown trout on Spring Creek in Pennsylvania is almost invisible.

currents. Once I'd spotted it, it became almost obvious, and I wondered why I hadn't seen it sooner, without the binoculars. But that is the nature of camouflage: concealing the obvious.

Trout colors change so they blend better with their environment. The change might be slight and is never abrupt, but taken from a dark environment, they will be darker; taken over lighter substrate, they will be lighter. That is why a trout's beauty fades if you catch it and kill it: They are not so pretty dead, because their colors slowly subside when life flows away from them.

Rooted plant beds are excellent concealment for trout and offer protection from predators. Trout can rest among all the channels the currents cut through the plants, down on the bottom and out of sight. When feeding higher in the water column, above the plants, all they have to do is slip back into them when something like the shadow of a bird or the flash of a fly line sets off their alarms.

Ledges, trenches, and submerged rocks or bigger boulders are the most common types of protection from predators in most trout streams. It's always a surprise to see how effectively trout can use the smallest of hiding places, especially if the lie offers even the slightest patch of shade. When threat lurks, alarmed trout tuck themselves in and barely quiver a fin. They become extensions of this rock or that bit of ledge. Even when not alarmed, they take similar hidden positions from which they can safely survey a nurturing current.

I recently fished a medium-size trout stream in an open ponderosa pine forest. It was September, the banks were grassy, and big grasshoppers leaped into the air at my every step. The sun was high and bright, so I focused my fishing on deeper central currents, on the assumption that trout would be hanging out where they had at least slight concealment from overhead predation. The water was too thin at the edges, in my estimation, to hold any but tiddler trout. So I waded the shallows to reach the deeper water toward the center of the stream. The streambed was an ankle sprainer, with lots of boulders the size of basketballs and larger.

Surprisingly few trout made attempts at my dry fly, naturally a hopper pattern, in the deeper water where I suspected they'd be. I'd fished nearly an hour and had covered about a quarter mile of stream, wondering what I was doing wrong, before it occurred to me that I was pushing more than occasional trout ahead of me as I waded the shallows, and none of them were tiny. Every one of them bolted from water no more than a foot and a half deep, and never did I see one that emerged from anything but a black spot of shade from one of those boulders that caused me to stumble every few steps.

I redirected my casts directly upstream, to the shallow water ahead of me, a foot to at most five feet from the bank. The change in my luck was abrupt. It became even better when I began focusing my fishing not just on obvious holding lies along the bank, but also on those places where a boulder and the angle of the sun created a slight dark spot of shade. A nice native redside rainbow trout, twelve to sixteen inches long, arose from nearly every one of them.

When trout holding in water that is somewhat shallow are suddenly alarmed, they flee at once to what the late Charles Brooks, in *The Trout and the Stream*, called "bomb shelters." These are the deepest, darkest, and most remote parts of any piece of creek, stream, or river. Trout bury themselves far back in crevices between bottom rocks, they sink to the bottom of the deepest pool, they burrow to the backs of undercuts that can reach surprising distances beneath root balls and grassed banks. I'm not going to tell you to fish these shelters for the trout you've just driven into them. But it's worth mentioning that many shallow stretches of stream lack any such water nearby, and lacking the sanctuary, they will often lack trout as well, especially ones of any acceptable size.

It's far from true in every instance, but a stream with extensive shallows will often have good fishing only in water where trout have access to

When the sun is out and bright, especially when it's directly overhead, trout will seek shade if all else in that bit of stream is equal, and many times even if it's not. You can read the water by looking for shade and seeing if the water in the shade meets the other needs of trout. I have a favorite midsize stream that is hopper water. Trout, as they should, locate themselves along the edges of it, waiting for big terrestrials to make mistakes. But they don't line both edges. They can almost without variation be found on the bank that is cast into shade by the daily migration of the sun. I don't know if trout shuttle back and forth across big rivers to reach the shady side, but I am sure they do it on small streams and at least some midsize ones.

On larger rivers, it's more likely trout simply subside into the depths when the sun gets too bright for them. You might not be able to measure their return to activity by the slant of the sun, but trout often become willing to take dry flies along edge currents, even those in the sun, around four or five o'clock in the afternoon. It's not because the heat of the day is beginning to wane; it's because the light is less brutal and they're rising up out of their protective depths.

a bomb shelter. On occasion, I've fished long stretches of stream but caught trout only in water within fifty feet upstream or down from a deep pool. Of course, the deep pool itself nearly always holds trout. But it's also educational to find good fishing only in a tight relationship with the deep water, with long, barren stretches of empty water in between. You won't find such a situation often. You will not often be able to look at a stream and predict this precise situation. But learn to watch for areas where your success is limited to short stretches of water in which trout, given a fright, are able to dash for the depths. You might be able to focus on water that doesn't look much different from the long reaches above and below it but holds trout even though they are absent in the water that looks just as good.

TEMPERATURE AND OXYGEN

Temperature and oxygen, which satisfy the third need of trout, are intertwined factors. The colder the water, the more oxygen it entrains. High temperatures were once considered the primary factor in trout mortality during summer low-water conditions. But at high temperatures, water holds less oxygen, and it's now thought that the lack of oxygen in warm water, not high temperature itself, is most often the eventual and actual cause of death in these cases. The warmer the water, the faster the trout's metabolism and the more oxygen it requires. Ernest Schwiebert, in his thorough two-volume *Trout*, noted that "fish require four times as much oxygen at seventy-five degrees as they need at only forty-odd degrees." These two factors, oxygen and water temperature, are inseparable.

Water is a more stable medium than the atmosphere. Trout stream temperatures follow the mean atmospheric temperature by a few days; it takes water a bit longer to warm up toward air temperatures. This is why trout do not perish in sudden but short heat waves: The air gets hot faster than the water. The mean daily temperature of water is approximately the average of the daily high and nightly low. When you step into a trout stream to cool off on a hot day, you're enjoying some of last night's coolness that has not yet wicked off. Trout are saved by this same phenomenon: A trout stream simply does not get as warm as the noontime temperature of the hottest day.

Trout are most active in water through the wide range from 45°F to 65°F (8°C to 18°C). They will feed, sometimes eagerly, at temperatures of 40°F (5°C) and even below, but such activity hinges on the availability of food. Aquatic and terrestrial insects are not generally active at such low temperatures. A lot of trout activity is tied to insect activity. Most aquatic insects begin to emerge when temperatures inch into the upper 40s (8 to 10°C), and emergence peaks at temperatures between 50°F and 60°F (10°C and 15°C). Trout are most active at these temperatures because that is when they have the most to be active about.

But anglers may encounter exceptions. Just this January, Rick Hafele and I fished the Deschutes River with John Smeraglio, owner of a shop along the river and a guide there. We got into a hatch of *Baetis* mayflies with water temperatures measured at 40°F. A few trout rose to feed on them the first day, and we took several on dry flies. The second day, the hatch was stronger, with dozens of duns boating the currents at any given moment, but no trout that we were able to see ever responded to them. Air temperatures were much colder; we had ice in our guides both days but far more the second. And probably most of us have fished over midge hatches when the water was below 40°F, again with ice in our guides, though often with considerable success.

As temperatures rise, trout begin to suffer discomfort in water a bit warmer than 70°F (21°C). Brook trout and cutthroat cannot survive temperatures much higher than 75°F (24°C). Rainbow and brown trout can survive temperatures a few degrees higher than that, to just over 80°F (26°C) in rare circumstances where the water is highly charged with oxygen.

Sudden spikes in water temperature are most distressing to trout. Given time to adjust, they acclimate to normal seasonal high temperatures. If the change is abrupt, mortality might occur at temperatures a few degrees lower.

At the opposite end of the thermometer, low temperatures decrease the metabolism of trout, nearly halting their activity when it gets within a few degrees of the freezing point. But running water does not freeze except in northern latitudes that are beyond the range of what we normally consider trout waters, and most trout will survive so long as the water does not freeze. A rare season of anchor ice, which forms on the bottom and works its way upward, can devastate life in running water, killing aquatic nymphs and larvae. When it floats away from the bottom, it lifts part of the streambed with it. As the current tumbles loosened pieces of ice and rock along the bottom, they grind little lives to bits.

Water temperature can help you locate trout. When it's cold, trout naturally are found deep along the bottom, usually in slow to medium currents. When temperatures are in the more comfortable midranges, during which times insects are also most active, trout respond to the availability of food. If a hatch is happening, they'll be up and feeding on it. If no insects are available on the surface, trout will be found along the bottom, but they'll be on the lookout for food and therefore susceptible to your flies.

Whenever temperature and oxygen regimes approach uncomfortable zones, trout search for comfortable levels, just as they search for relief from strong currents and protection from predators. The outlet of a cold spring might provide a comfort zone when temperatures are high. The entrance of a fresh, cool tributary often gathers trout in hot weather. In very short streamcourses, trout might move toward the headwaters, up into

If the weather has been warm for a spell and you have trouble finding trout in all their normal lies, try prospecting around the confluence of any feeder stream you see entering the main river. Though they're not always cooler, as some feeders are more exposed to the sun than the rivers into which they flow, tributaries tend to be more forested and are always narrower, so they're more likely to be shaded. Many, such as the small Boulder Creek shown here entering the larger Lochsa River in Idaho, arrive tumbling over bouldered beds, assuring that they'll contain more oxygen than the main river. Trout sometimes pod in the nearest good holding lies downstream from the confluence, though this is more likely to happen in a forested or pastoral foothill stream than it is in a river like the Lochsa, rushing off the Rockies.

those tributaries, seeking cooler water. In many cases, they can do no more than move to where the water best suits their needs, which is why when temperatures are distressing, you will find most trout in and just downstream from riffles and rapids that are charged with oxygen.

I once searched the course of a familiar stream for most of a scorched midsummer day but couldn't locate its trout. I'd been having good fishing on that forested valley stream for weeks, drawing trout up to dry flies in all the long runs and slow pools. But suddenly they were no longer in water where I'd consistently been finding them. I sat down to rest in the shade for a while in early afternoon and then got up to meander back downstream to the rig, intending to give up.

I tied on a wet fly to dabble here and there as I went. A wet fly fished on the swing through all the odds and ends of water where you'd not normally toss a fly is one of the best ways to find trout in water where they suddenly seem to be absent. Among the places I casually fished was a short, white set of rapids that fed into the choppy top end of a riffle. No fish had ever held there when I fished it before; I'd tried it often enough that summer without catching anything to have given up on it and therefore bypassed it on my way upstream that morning. But this time a trout climbed on the first cast, and I caught two more there before moving on to the next whitewater. I landed half a dozen nice trout in that same type of well-oxygenated water before I got back to the rig and quit.

A bit of a warning is appropriate here. Those trout were all charged with oxygen and vigorous, or I'd have ceased fishing for them. If you manage to find trout during a warm spell, in water that approaches the upper limits of their ability to survive, and they're willing to take your flies but fail to put up much of a fight, it's an indication that they're already stressed. If you hook them and play them to your hand or net, then are forced to revive them for a long time before releasing them, it's likely they'll never recover, even if they appear strong when they swim away. Leave off fishing for them and return to catch them another day, when water temperatures have returned to more normal levels.

FOOD

The fourth basic need of trout—food—often overrides all their other needs. This need for nutrition operates on some basic formulas. The first and perhaps most important: *The energy gained from a bit of food must exceed the energy expended in the effort to acquire it.* If a large trout were to dash its bulk through several feet of water, battling a strong current to do it, then leap high into the air merely to take a size 24 gnat, it would lose calories, and therefore weight, in the bargain. This might be the sound basis for a weight-loss diet, but it's well known that trout, lacking vanity, seldom glance into a mirror and gasp in fright because they're fat.

This rule of adequate return is obeyed more religiously as a trout gets larger. It's common to see a troutkin fling itself into the air after a small and elusive aerial gnat or microcaddis. Even then it might be gaining more than it loses, because it takes far less energy to propel the small trout, and a tiny bite might be a better bargain for it. You will rarely see a trout of substantial size make the same mistake, though you'll sometimes see a big fish leap awkwardly to take a large insect such as a cranefly, fall caddis, or salmon fly adult out of the air.

In the presence of an abundance of food, trout will rise up to the top, expose themselves, and risk predation. It's not common, but also not entirely rare, to see an osprey explode into the midst of a pod of trout feeding on a hatch of insects, take its tithe of them, and put the fish down for a brief time, only to see the trout come back up and start feeding again soon. They'll feed as long as the insects are present, in that everlasting measure of one need against the other: Getting fed comes first for trout and will be done at the risk of being fed upon themselves.

A small trout will often appear to hold in currents that seem unlikely to deliver more food than will fuel its fight against them. It might have compromised for a break-even deal: The trout exists in equilibrium, taking in as much energy as it expends, without growing. But it's much more likely that the trout has found some crease in the current that allows it to conserve energy or is taking in far more nutrition from tiny bites than we are able to notice from our view outside the water. Trout are like kids: If we could ever interview one, it would probably fold its fins importantly across its chest and announce that its greatest goal in life is to eat lots and grow up to be big and strong.

Truly large trout rarely move far for a single insect. But they do take up stations and feed rhythmically when a hatch is on, when a little energy can be expended to take in a succession of bites, even very small ones, without a lot of wasted movement. It's far from uncommon to see four- and five-pound trout holding high in the water and feeding eagerly on size 22 or even 24 Tricos or midges, as long as the insects are so abundant that they can be sipped one after another. A big trout will also chase a sculpin or baitfish down to its death, clear across the river, considering that the effort will be well rewarded if it manages to catch it.

Conservation of energy is the first rule of feeding fish. The second is this: *When food is abundant and easy to get, trout often neglect the need for protection from predators in order to feed.* During a heavy hatch of insects, trout venture out of more protective lies, hold close to the surface, and expose themselves to all sorts of danger, including that presented by you and me. They do this even at the price of a tithe of themselves. An example is the pods of trout that work flats of famous spring creeks and tailwaters, risking both anglers and the occasional passing osprey. When a hatch is heavy enough, trout will feed disdainfully almost at an angler's feet, turning selective wet noses away from all but the most exact imitations and most careful presentations. An occasional pod-fellow sacrificed to an osprey doesn't even stop their feeding.

The larger the trout, the more wary it will be when exposed to danger. It's difficult to say whether a trout is wary because it is old or old because it is wary. But the largest fish are least likely to expose themselves to potential danger and most likely to be both cautious and superselective when they do. It's true that the biggest trout on the flats of famous waters, such as the Railroad Ranch stretch of the Henrys Fork of the Snake in Idaho or the Letort in Pennsylvania, feed on small insects just like the rest of the trout. But they usually do it in the most difficult water, where they are protected by rooted plant beds or fallen jack pines.

One must also consider the possibility that the very largest trout might abandon feeding on dangerous flats. Trout that outweigh their placid-water neighbors by three or four pounds are taken every year in the turbulent and canyoned water just upstream and down from the Henrys Fork flats. And there are rumors of big browns on Pennsylvania's Spruce Run that hide most of the time, ignore insect activity, and come out to eat their neighbor trout at night.

TERRITORY

This subject of neighborliness brings up its opposite: the territorial instinct of trout. Fry and fingerlings hover in schools in still backwaters along the edges of moving water. But in running water, they keep in contact with their brethren while aggressively defending their own bit of bottom within the school. As they grow larger, their territories expand. As Schwiebert pointed out in *Trout*, "The prime fish will establish territoriality over the best hiding and feeding places." Reading water is a study in finding the best places, as it is there that the biggest fish will be found.

Trout are territorial in relation to the speed of the current in which they're found. In a broad, brisk, and unfeatured riffle like this one on the Yellowstone River below Big Timber, Montana, trout are sprinkled on separate lies, wherever they find a boulder or depression to deflect the current. You won't find them in pods, and you'll have to use your fly pattern to read the water, casting over all of it in a disciplined search for scattered lies that are not marked on the surface. After you catch one fish, repeat the cast and fish the lie out a bit more, but it's likely you've caught the only trout in it. Move on fairly quickly, continuing to explore the water for those hidden lies.

Different trout species exhibit varying degrees of aggression in establishing and defending territories. Where species overlap, browns are the most aggressive, rainbows next, and cutthroats a bit meeker, while brookies often get chased into the neglected headwaters of their own native streams by more aggressive species that were stocked later. The result can be that a smaller fish of one species displaces a larger fish of the other. Habitat preferences work into this equation, however, so it's doubtful that it's common, at least in natural situations, for a small rainbow trout to chase out a larger cutthroat, simply because a rainbow would prefer faster water, a cutthroat a more placid pool.

But there are far fewer natural situations left than there used to be. Most trout streams have been stocked. Any American stream that holds brown trout has been stocked, because browns are not native to this continent. Any midwestern or eastern stream that holds rainbow trout has been stocked, because that species is not native anywhere east of the Rockies. So displacement by stocking, and later by wild reproduction of non-native species, becomes a large factor.

It's very distressing when hatchery trout are suddenly dumped into a stream that already holds a population of native trout. Territory is piffle to hatchery fish. Having lived in holding ponds, they are accustomed to the constant fin-tip nearness of other trout. When a gob of them is injected into a stream, they mill about with no idea where to be nor fear of other trout, and the result is confusion among native trout that were already there. These unhappy fish can't drive out all the intruders, nor can they stand to share their territories with them, so they move out, leaving the field to the inferior fish. When displaced from their territories, native trout are distressed, frequently can't find enough food to survive, and as a consequence often die. It leaves one wondering how stocking trout in healthy streams can be considered anything but harmful to our fishing. Clearly it's harmful to native trout.

Creeks and streams and finally rivers get larger as they flow downstream, gathering more water. Trout drop downstream as they grow larger, seeking ever larger territories. According to Hynes in *The Ecology of Running Waters*, trout "are limited at the downstream end only by the suitability of the available territories." That is why many trophy trout, especially piscivorous browns, are caught in the downstream stretches of big rivers that are giving way from being trout waters to becoming warm-water fisheries. It also explains why an occasional lunker trout is written up in the newspapers, taken by some barefoot innocent dangling a worm in a muddied river with the mere intention of catching some crappie or carp from water that is known to hold trout, but only far upstream where it's prettier.

The best territories meet the most needs of the trout. W. B. Willers stated in *Trout Biology* that "the territory is essentially a feeding area, and within it there is, according to most accounts, a single spot, the *station*, from which the territory is defended and feeding excursions are made." The station will be a place that offers shelter from currents, protection from predators, and a clear opportunity at sufficient food.

A trout spends most of its time living within its territory, holding at its station, facing upstream. From this point, it has a view of its estate. It can dash out to defend its territory from an intruder. It can nose out to accept any bit of food delivered downstream to it by the current.

It's vital to note that the instinct to defend a territory exists only in moving water. Willers noted that "strong territoriality . . . virtually disappears where there's no current. As water velocity decreases and finally approaches zero, mature fish begin to move about in a random way."

This is an exaggerated photo of what you're reading water to find in most of your trout fishing: a single trout holding its station in moving water. You'll rarely find one so evident to the eye. Most of the time you'll have to read the shape of the water to locate such lies, and then fish them without direct evidence that a trout is down there. But trout always have a reason for holding where they do, and your job is to read the surface and the currents, looking for signs of those likely lies.

Several interesting bits of trout behavior stem from territoriality. It's the reason you'll usually take the biggest trout on your first few casts when you fish pocket water. The dominant trout defends the small window on the surface that makes this station the best lie. When something lands up there, that big trout is the first to see it and rush to get it. When you learn to read the water right, you can often pop your dry fly or plunk your nymph right to the water where the biggest trout lives.

Territoriality also explains why the trout you take from a riffle are scattered across it, rather than bunched up in a single area. In fast-moving water, each trout establishes a small territory, usually with a station in front of or behind some object that deflects the current. Unless these potential territories are concentrated in a relatively small area, trout will be sprinkled around a riffle like eggs on Easter morning. When you move from the riffle downstream to a slower run, relatively still flat, or pool, you will find trout working in pods. The water has slowed, their territorial instinct has decreased, and they're willing to gang up to feed.

Territory is not a factor in bomb shelters. If trout can be said to worry, when crowded into a sanctuary they're too worried about outside danger to calculate their distance from one another.

SPAWNING GRAVEL AND REARING HABITAT

Spawning gravel and rearing habitat are not survival needs for *individual* adult trout. They do not predict where you'll find trout, except during the spawning season. But they are requirements for a self-sustaining trout *population*. Streams that lack sufficient natural spawning must be augmented by stocking if they are to offer good fishing. This need for stocking is especially tragic where humans have caused damage to spawning beds in streams that have lots of feed for trout and at one time had plenty of good gravel for its native trout population.

Absence of adequate recruitment can sometimes be detected by an unnatural lack of smaller trout. While I was away in the army, loggers moved into the watershed of my favorite stream near home, in the forested coastal hills of Oregon, and knocked all its trees to the ground. The stream's spawning beds were silted in, destroyed. I had good fishing for the first couple years after I returned, employing strict catch-and-release. But the numbers of juvenile fish in the four- to eight-inch range dwindled rapidly. Before long, there were no fish left except ten- to twelve-inchers. It was a very small stream, and these were very big fish for it.

Individual trout did remarkably well. They were fat and well fed. At first there were three or four in a modest pool, then in a few seasons just one or two. When these older trout died off, the stream also turned its belly to the sky. There were almost no smaller fish to move into its open territories. Enough native trout will likely remain to seed the stream, but the population won't recover until the forest has regenerated enough to prevent spates from delivering more silt to the spawning beds. It will be decades before those spawning beds are clean and recruitment of native trout becomes even close to what it once was. Pressure to stock the stream with non-native trout would be intense were it not so remote and difficult that few others have any interest in fishing it.

That was once my home stream, and its loss propelled me to explore and fish lots of other rivers. In order to catch trout in them, I had to find out where those trout held. In order to do that, I had to learn to read trout water.

Reading water is a matter of learning what is comfort to a trout, then discovering how the stream provides that comfort. A trout is happy, if trout can be said to experience happiness, when its needs are met: when it has shelter from the current, protection from predators, and a temperature and oxygen regime in which it can thrive. It must have a territory that provides food sufficient for its size. A bomb shelter nearby increases the value of a territory to a fish and helps attract a larger trout and keep it present.

When a trout has found its ideal territory, it will be found most often holding on its station somewhere inside that territory. It will maintain an upstream posture, from which position it can survey its estate, defending it against intruders and picking off what food it offers. It will spend most of its time in a combination of holding and feeding.

The key to finding trout is acquiring the knowledge that lets you recognize lies where the stream meets their needs and therefore makes them comfortable. The central ability to catch more trout lies in learning to locate territories sufficient to attract them and keep them content, and then pinpointing your casts to the most likely stations in those territories. The secret to catching larger trout lies in learning to recognize water that meets trout needs with such abundance that it lets them prosper without much effort so that they can gain great size.

2

Structure of a Stream

The structure of a stream dictates where and how it will meet the needs of trout and is of more than minor interest to a successful trout fisherman.

Running water is divided by hydrologists into a simple set of classifications. The water types are riffles, runs, pools, flats, rapids, and cascades. That nearly translates into the way anglers think of rivers, though rapids and cascades are fishable only in pockets, so we tend to excise pocket water out of rapids and cascades and consider it a water type on its own, ignoring the frothed and fishless water all around it. And we add a final type, bank water, because lots of trout are caught along banks, and it's fished differently than the other water types.

One note belongs here at the beginning: Though streams are divided simply into water types on paper, they're not so thoughtfully constructed when you go out and get wet in them. Riffles merge into runs, and the border between is usually, but not always, distinct. Runs flow into pools; exactly where they cease to be one and become the other isn't always clear. Flat water is a flat unless it's a bit deeper and faster, in which case it is defined as a run. But most of the water in a creek, stream, or river falls neatly into one of the classifications, and water that doesn't is usually a brief gradient from one type to another.

You'll be fishing with certain tackle and a specific rig when you reach transitional water. It's doubtful that you'll want to gallop back to your vehicle to trade your 8-foot rod for a 9-footer, your dry line for a wet-tip, and even reconstruct your rigging from a dry fly to a nymph and indicator just so you can fish a few feet of water that is no longer a riffle but not yet a run.

As you move from one water type through transitional water to another water type, keep doing what you were doing until it no longer works. Then figure out what to do based on what the new water type tells you about where its fish hold, how they are feeding or not feeding, and how you should rig to catch them.

Though you'll rarely want to hike as high as I have here above Beavertail Campground on my home Deschutes River, getting an overview of any river will help you learn to read its water and locate its trout. You'll be able to spot the riffles and runs, the pools and flats where trout might be found rising to insects, and the bank water where they'll tuck in tight but be susceptible to your caddis, salmon fly, or grasshopper dry flies. An overview of any creek, stream, or river, especially one to which you're new, will instruct you about its shape, structure, and water types that are most likely to be productive. Not all trout waters allow such a look, but if you can get a bit of a view of whatever water you're on, you'll understand it better and get a better sense of how to read it and find its trout.

WATER TYPES

Though they are defined in more detail in later chapters, some early notes on water types will be helpful here because the rest of this book is constructed around them.

Water type is a function of three things: the gradient, or rate of descent, of the stream bottom; the type of material that lies on the bottom; and the depth of the water that flows over it. Each water type has a different structure, and each structure pleases trout in a different way.

Riffles are stretches of stream that are cobbled on the bottom and shallow enough that this bumpiness is conveyed up to the surface. What you see when you approach a riffle is a surface that is rippled or choppy, depending on the size of the stones on the bottom and the depth of the water between the bottom and top. Normally a riffle's rocks are softball-size or smaller, its water three feet deep or shallower. The streambed is at least slightly tilted, so the water moves at an accelerated pace.

Some riffles are tipped so steeply that the water rushes over them too fast and is too shallow to hold trout. They can't fight the current and are exposed to overhead predation. These riffles aren't worth fishing, but they are among the richest areas on any stream in terms of aquatic insect life. You will almost always find trout holding in the nearest slower and deeper water downstream from such a riffle, feeding greedily on aquatic insects that lose their grip in the fast water and get delivered down the riffles.

Runs are moderate to deep water flowing evenly over bottoms that tend to larger rocks, bedrock and ledges, or sand and silt. The gradient is not so steep as in a riffle, and the water is not in such a hurry, though it still flows forcefully. Typical trout runs are between three and six feet deep, though some are shallower but not defined as riffles because they do not have cobbled bottoms and choppy surfaces, and some are deeper but not defined as pools because their depths and current speeds are relatively even from the head to end of the run.

The surface of a run is often unbroken, fairly smooth on top, but many runs that are just three to five feet deep carry the choppiness of the riffles upstream from them downstream through their entire lengths. Almost all runs display some unruliness where boulders break the surface or any kind of obstruction is so close to the surface that boils or swirls swim up to the top. But a run does not have the even, cobbled bottom of a riffle, and its bottom structure is usually reflected in a relatively calm surface with patches of broken water, rather than the consistent chop of a riffle.

Pools are places where the streambed dips, then rises again. They are gouges or bowls eroded into the bottom and are deeper than the water

As you move along any creek, stream, or river, you'll often find that trout are holding most consistently in one type of water, in this case the run just downstream from a rich riffle on Montana's Rock Creek. Ascertaining which water type is most productive, by fishing all the water types and noticing where you catch trout, is a form of reading trout water. You can then focus your fishing on that water type as you travel along and thereby increase your take. This can be put to use on any size stream, but it's especially effective when you're in a boat, on a long float of a river that offers you lots of options in terms of water types. Once you've figured out which one provides the most action, you can focus on it, spending the bulk of your time fishing that type of water and saving time not fishing what might at the moment be empty water.

Never neglect all that other water. In the first place, it's rarely entirely empty, and it always has pinpoint lies that are likely to provide trout even if the water all around the lie isn't producing. And the moment can change. When the sun begins to slant down, trout along banks that have been unproductive move up and might like a chance to whack your flies; insects quit tumbling down from that riffle upstream and trout in the shallow run move out into the safer depths; or trout in the deep pool you've been probing move to feed on an evening hatch that suddenly blossoms on the tailout.

The surface of any moving water reflects the structure of its bottom, though the deeper and slower the water, the less that reflection reaches the top and the more difficult it is to read. In this case, on California's Pit River, it's easy to see that the bottom is studded with big boulders. Some send visible signs up to the surface, but most do not. That means you'd want to fish all the water across this bench, because trout might be anywhere down there. But that long, narrow line of slick water on the surface in front of me, downstream from the visible boulder on the shelf, is an obvious place where the current is broken by the structure of the stream, and it's the most likely place to find a big trout or a string of them lined up the length of that seam. I'm positioning myself to fish it with short casts, nymph and indicator, and a high-sticked rod.

upstream or down from them. Pools are the deepest parts of any stream, but depth is always relative to what's around it. A plunge pool in a mountain creek might be just three feet deep. A pool that punctuates a major river is more likely to be ten to fifteen feet deep.

A pool is a hesitation in the water's way toward the ocean. It is slow; sometimes it nearly stops. Where a current tongue enters from a riffle at the head of a pool, the water is rough on top, with eddies off to the sides. But the body and tailout of a pool are smooth on the surface. The bottom in the deep center of a pool tends to be the finest pebbles, gravel, and sediment in the stream, sometimes studded by large features such as boulders or even sunken logs.

Flats are about the same depth as riffles and shallow runs, but they are not so steep, their water is not so fast, and their bottoms are composed of sand and silt or small pebbles and gravel. The smoothness of the bottom is reflected on the surface, which is typically slick. From a distance, flats appear to have absolutely even flows, with no features or conflicting currents, but they often have minicurrents that push and pull a leader all around, causing drag on your dry fly that you will never notice but trout will never ignore.

Flats as we define them in fishing terms are almost always found in streams with stabilized flows, typically spring creeks or tailwaters, though in rare cases such as the Yellowstone River downstream from the lake, flows are stabilized by still-water origins. Streams subject to spates do not allow the deposition of sediment, sand, and finer pebbles and stones. The smaller material is washed out annually, the bottom becomes cobble and rock or even boulders, and what would be a flat is instead rough on top and defined as either a riffle or a run. On streams with stable flows, sediment is deposited, not washed away, and the surface, reflecting the streambed, is smooth on top, which is why we call it a flat.

The water in flats varies from a foot deep to about four or five feet, though that's a deep one. Because the current is slow and the bottom often silted, vegetation gets a chance to take root. Plant beds become a major—almost a defining—feature of flats. Plant beds are rich in aquatic insects, and most flats are therefore rich in trout, which often feed selectively on a specific stage of a single species. Flats can be the most challenging, the most damnably difficult, places to fish on any stream.

Pocket water is located in rapids and less often in cascades. Rapids are sudden steep tiltings of the streambed over bottoms of rocks and boulders or ledgerock and bedrock, creating turbulent whitewater. Cascades are rapids strung together, with an even faster drop and current that is usually too brutal for trout to survive except where it's broken.

Pockets are created in such fast water wherever an obstruction is large enough to cause the current to eddy in behind it. Trout can hold in comfort in these small eddies, with all their needs met. They find shelter from the current, protection from predators, and the constant delivery of food caught in the turbulent drift. The size of the pocket dictates the size of the territory, and therefore the size of the trout that will hold in the pocket. A large pocket might hold a string of trout lined up in the calmed flows downstream from an obstruction to the current. A soft pillow of calm water also builds up in front of a boulder, breaking brutal flows. One or sometimes two trout will hold in that soft water if it forms a pocket large enough for them to avoid fighting the current.

Bank water is the last water type that we anglers tend to separate out in moving water. It's given its own later chapter, because a certain narrow set of circumstances determines whether bank water will hold trout or lack them, and the criteria are the same whether it's the bank of a riffle, run, pool, or flat. Most trout streams have at least some fishable reaches of bank water, others have an abundance of it, but most bank water along typical trout streams is low-percentage water. Learning to read it right, to separate good bank water from bad, will often put you onto trout that other folks overlook and also let you know to ignore a lot of water that other people waste hours fishing fruitlessly.

STREAM TYPES

I'd like to define three words—*creek*, *stream*, and *river*—though we all know vaguely what they mean. They all hinge on *stream*, because a stream, according to my old Webster's, is any flow of water or liquid, which means both your coffee spill and the Mississippi River. A *creek* is defined as a small stream, and a *river* is defined as a stream larger than a creek. That's about all the dictionary gives us to go on. All three are defined in terms of each other, and the definitions are therefore relatively meaningless. The cross references become more confused when the same dictionary also calls a stream a small river.

We can work with this, but we need to pull out definitions that mean something in our angling terms. To begin with, a creek and a river are both correctly called streams: A creek is a small stream, a river is a large stream. When the word *stream* is used more narrowly to describe the size of a piece of trout water, in the line of succession from creek to stream to river, we can accept Webster's definition as a small river, which I interpret to mean the typical medium-size trout streams that most of us fish most of the time.

This might seem a semantic tangle, but the result is both simple and useful: A creek is small, a stream is medium, and a river is large. I like to think in terms of fly casts. A creek can easily be covered from any position, and from side to side, with a short cast, say twenty to twenty-five feet. A stream can be covered with modest casts, forty to fifty feet, though you'll sometimes have to wade to get into the precise right position to cover a specific bit of water correctly. A river can and usually should be covered with short to moderate casts, but sometimes long ones will serve you better, and you'll always need to wade from position to position to cover all of the water.

Creeks are often called microcosms of big rivers, in terms of reading water. Some are that, but most are not. They're small streams, each is different from all others, and you learn to read them in their own right. It's true that their water types are restricted in number and condensed in space, so they're often much easier to read than larger streams and rivers. They are therefore excellent places to learn to read water. The mountain plunge pool stream my daughter Kosumo is fishing writes its water types quite clearly: miniature waterfalls excavating minor depths that are perfect for feeding trout, sheltering them from currents and protecting them from predators.

Even more simply, a creek is a short cast across, a stream is a medium cast across, and a river is a long cast across or wider.

A *stream system* is the whole works, from the smallest headwater creek, through a medium-size trout stream, to where the flow of water finally bows and exits into an ocean as a large river. Water gathers in the hills and flows downstream, increasing in size as it goes. But a stream system constantly erodes its way upstream. This is part of the process that wears away mountain ranges, an operation you don't want to stand around and watch happen. A stream system is a succession of stream types, from first-order headwaters too small to hold trout, through mountain creeks, to typical trout streams, on downstream to large trout rivers, and in many cases beyond to warm-water rivers.

Cedar Run, in Pennsylvania, is a classic medium-size trout stream. Such streams come in a wide variety of types, always depending on the topography through which they descend. Some are mountain, some are foothill, and many are out in the flatlands. But stream sizes follow a logical progression, from creeks in the steeper headwater areas, to medium-size streams that have gathered two or three creeks and flow in the foothills, to larger rivers that have accumulated creeks and streams, reached maturity, and eroded their own courses to at least the relative flatness of river valleys. There are a vast number of exceptions, but the typical medium-size trout stream displays the riffle-run-pool structure that we think of as defining a classic trout stream.

Even many of those streams have sections that tip down and become bounding, with more pocket water than any other water type, and other sections that level out, meander, and offer glides, gentle bend pools, and undercut bank water. A stream is always defined by the amount of water it escorts along—creeks, streams, and rivers—and the type of watershed through which it flows: mountain, foothill, or valley. But you'll find creeks in valleys and rivers in mountains. In general, though, water shapes the land through which it flows: Creeks plunge off mountains, streams drain foothills, and rivers take to valleys.

Pine Creek, in Pennsylvania, is an example of a mature river. Like a classic stream, it has riffles and runs and pools, all writ large. But the Delaware River, Montana's Madison and Big Hole, Idaho's Henrys Fork of the Snake, and Oregon's Deschutes are also big rivers, and all are very different. If you tried to read them all the same way, you'd be very disappointed. You read big rivers by breaking them down into their parts—riffles, runs, pools, flats, bank water, in some cases even pocket water—and then read them a part at a time. A riffle in a big river might be bigger and broader, but it is not much different from a riffle in a classic trout stream. All the other water types are read the same way: Look for the places where the water takes care of its trout, and that's where you'll find them, the same as you would in creeks and midsize trout streams.

In a *mountain creek*, trout hold mostly in pools or pockets because everything else is usually rapids or a cascade. In trout streams, which have gathered from two to a few creeks to them, trout find the widest diversity of habitats. These streams display the classic riffle-run-pool structure that we all love to fish. In large trout rivers, good lies are farther apart and less well defined, but they are still riffles and runs and pools and flats, plus bank water. They often hold pods of trout and are also likely to hold the largest trout. Eventually the stream system descends to the flat lowlands, where it frequently turns into a warm-water fishery.

Meadow streams, many of which arise as spring creeks, are variations in the simple succession of stream types outlined above. They range in size

Meadow and Freestone Streams

Although streams are often classified as either *spring creek* or *freestone*, one is a description of a source and the other is a definition of a bottom type. If the term *spring creek* is used to define one type of stream, then *spate* should describe the other. They are both sources; one springs out of the ground, the other descends from the sky as rain and seeps to the stream as groundwater or rushes to it as runoff. It gets more confusing when you realize that spring creeks have tumbling freestone reaches, and spate streams flatten out in peaceful, meandering meadow reaches.

It's much more meaningful to describe flowing water as either *meadow* or *freestone*, because both describe structure rather than source. A meadow stretch of stream—whether its source is in springs, as for Pennsylvania's Falling Springs shown here, or rain, as it might be for a mountain stream crossing a level bench—will reflect the gentle gradient of its watershed. It will have slow flows, meanders, and the deposition of fine gravel and sediments in its bed.

A freestone section of a creek, stream, or river, no matter what its source, will be shaped by its steeper gradient. It will be faster; cut a more direct course, bouncing back and forth between banks that constrict it, rather than meandering; and have a bottom composed of much larger stones and boulders. The west fork of Rock Creek near Red Lodge, Montana, flows out of the mountainous country north of Yellowstone Park. If it's not typical for a freestone stream, that's because such streams come in as many shapes as their size and the steepness of the terrain through which they flow allow.

from tiny springheads that you can jump across to broad flats in large rivers, such as the Railroad Ranch mileage on the Henrys Fork of the Snake; this famous flat is fed from two sources, one a spring creek and the other a tailwater.

Tailwaters below man-made dams are another variation on the creek-stream-river succession. These are a relatively new type of fishery. Some of them are short-lived, depending on a set of circumstances that blooms into great trout fishing, then bursts and leaves it bad. Many others seem to be lasting fisheries, often created in waters that were warm-water reaches before the dams were built. Because some of the very finest trout fishing currently crops up below dams, tailwaters are worth a careful look in a later chapter.

Each of the water types, and each of these stream types, will be covered in separate chapters. For now, let's look at some features that affect all stream types.

Gradient is a measure of a stream's steepness. This in turn defines the water types it will offer its trout. A stream system works its flatness, in geographic time, toward its headwaters. Its gradient is steepest at the highest end, more modest in the middle reaches, and flattens out in the lower regions, where it becomes a mature river. A stream system's gradient is closely related to the geography of the land through which it flows. A creek plunging steeply down the slopes of a mountain is not at all like a foothill trout stream or a sleepy river that drains prairie country.

When a stream drops with considerable speed, it carries away fine material and erodes its way to stones, boulders, and bedrock. The term *freestone* refers to any stream that has a typical riffle-run-pool structure and a bottom composed of rock. The faster a stream falls, the larger the materials that form its features. One of my favorite local creeks plunges over and around boulders that run from baseball-size up to some you could park a car in if they had garage doors. The average trout stream, lower down in a system, has a gentler gradient and a bottom composed of the widest variety of material, from sand, pebbles, and cobble to rocks, rubble, and scattered boulders. The lower reaches of rivers have the least steep gradients and accumulate the finest bottom materials. When a river has reached the point where its bottom is composed almost entirely of sand and silt, it is also often getting beyond the point where it is considered favorable habitat by many trout because of seasonal high temperatures that exceed a trout's ability to survive.

Freestone streams have wide seasonal variations in flow. In late winter and spring, they are full and strong from snowmelt and rain. These high flows tear out vegetation and erode the banks far back, in most places in the shape of sloped gravel bars on the insides of bends, with deeper water pushed to the outside. In summer and fall, freestone stream flows drop, usually leaving broad gravel bars along the inside edges of their curves, with the deeper fishable water against the far bank.

Meadow streams, as opposed to freestone streams, have gentle gradients and tend to meander. Seasonal flows are at least reasonably stable, taming spates and reducing scour. Meadow stream bottoms are commonly composed of sand, silt, and pebbles, often with rooted plant beds. There is less erosion at the edges, and the banks crowd in against the stream and drop off steeply into relatively deep water, often creating dark undercuts that make excellent lies for trout.

Neither rooted nor attached aquatic vegetation—that growing on streambed stones—can get a firm grip on the bottom in strong currents. It does not survive in waters subject to winter and spring scour. Vegetation growth is characteristic of meadow streams and spring creeks; it is almost absent in all but the quietest reaches of freestone streams. Because a dam tames a tailwater and protects it from scour, a rich bloom of vegetation, and therefore a sudden increase in the abundance of aquatic insects and crustaceans, is generally one of the major reasons for the boom of such a fishery.

We tend to think of streams as either freestone or meadow types, wanting them, like we want most things, to fit into neat categories. But many streams are combinations, with some reaches of both freestone and meadow types. The Gibbon River in Yellowstone Park is typical of this kind of structure, offering lots of excellent water of both types, meadow and freestone.

The terms *meadow stream* and *spring creek* do not mean the same thing, though most fishermen tend to mean the same thing when they use either term. A spring creek arises from springhead sources, but the resultant stream often flows in and out of both freestone and meadow reaches. On the other side of the same coin, it's common for freestone streams, with no stable spring sources, to surprise us with beautiful meadow stream stretches. Few streams are all of one type and none of the other.

STREAM ECOLOGY

The ecology of a watershed has everything to do with what kind of stream it sponsors. In timbered watersheds, shade and the constant release of cool groundwater stabilize flows and regulate temperatures. In most forested areas, the trees along the edge of a stream are deciduous, enriching the water annually with leaf fall. Decomposing leaves become energy for bacteria, which become energy for aquatic insects, which become energy for trout.

Desert watersheds offer less defense for their streams. Their flows are marginally protected by narrow bands of willows and shade trees in the river bottoms. But the upland area of a desert watershed releases water quickly. Overgrazing is almost universal now in sagebrush country, and the affinity of cattle for water has caused many river bottom-lands to be the first areas stripped. In typical desert streams today, flows fluctuate widely and temperatures can be brutal, with abrupt daily and punishing annual swings. It's no accident that much of the best trout fishing in arid lands is in tailwaters that have reduced and stabilized stream temperatures.

Many streams have pastoral watersheds. These flow through farm country and are some of our most beautiful trout waters. The type of farming, the way it is done, and the respect landowners have for the streams all influence the condition of the water and the quality of the angling it offers.

The condition of a watershed directly determines the quality of the stream that drains it. Streams are affected by all things that happen on the land around them. In the forested Northwest, logging is the primary killer of streams. Carelessly done, it causes erosion, silting of the streambed, clogging of the spawning beds, and a reduction of insect populations that trout eat.

Overgrazing on desert watersheds, cropping grasses down to the soil, limits the land's ability to impede the rush of water into the streambed after a cloudburst. Cattle trample banks and cause silting of stream bottoms, limiting the streambed's chance of withstanding damage. Lack of shade causes overheating.

Mining is a less frequent problem but a more damaging one wherever it occurs without thought for nearby streams. Chemical leaching from tailings, if not neutralized, can sterilize streams, in some cases even painting their bottom rocks red.

In the industrialized states, the biggest problems are manufacturing plants in watersheds and development that sometimes almost displaces a stream. In some cases, trout streams have been turned into channelized ditches. Trout fishermen generally view cement viaducts as degraded watersheds.

Riparian zones, the narrow strips of watershed abutting streams, are being developed everywhere at a constantly increasing rate. Subdivisions and housing developments are crowding in wherever they can, to get a view of what will no longer be worth viewing when the subdivision is done.

In Japan, which I sometimes refer to as "the land of cement," stream structure can be very strange. It can also be beautiful, and if the streams are not productive, it's not because they're not pristine, but because catch-and-release is rarely practiced there. It's easy to forget that the Japanese have been shaping their streams for a few thousand years—though not all of those years with cement—and the streams, at least in the mountains where human populations are not dense, have worked the shapings of man into nature. The trout do fine if they're not caught and killed. Sometimes in Japan, you read water by reading man-made structures, such as cement blocks that have been placed to impede brutal spate flows but also serve to shelter trout from those same torrents.

WATER QUALITY

Water quality is a result of the things that have or have not been done to a watershed and the riparian zone. It's impossible to separate water quality from the quality of the watershed from which a stream arises.

Water chemistry is the first factor to consider in water quality. Running water is a richer environment than still water because the flow constantly delivers a fresh supply of nutrients to aquatic life. Chemicals that are the building blocks of life—nitrogen, calcium, carbon, oxygen, and trace elements—are suspended in the water and brought right to the plants.

Acidity is a major factor in a stream's ability to provide a trout fishery. The more acidic the water—the lower its pH—the less fertile it is. The higher the pH—basically, the more calcium it contains—the richer it is. Rainwater and groundwater are more acidic than springwater, especially where this last leaches through highly calcified soils. Spring creeks in limestone country carry the most usable nutrients for plant life, resulting in their richness in all the links of the food chain that lead to trout.

Pollution is a large factor in water quality. It's almost always a negative factor, but in rare cases it can enrich a river, causing plant life and therefore aquatic insects to flourish in what were previously less fertile waters. An example of this at one time was the rich fishery that existed below Calgary on the Bow River in Alberta, while fishing in the miles of water above the city was not as good. The improvement in the fishery below the city was based on the arrival of nutrients from Calgary's sewage treatment plants. They have been cleaned up, and for some years the fishery declined. Now it has returned to what it once was, though its fine quality is no longer based on nutrients from sewage.

Turbidity is a primary factor in water quality. Water that entrains a large load of silt impedes light, cuts down photosynthetic plant growth, and reduces the vegetative base on which the life of the stream thrives. Turbid water also deposits silt, clogging the bottom of the stream, limiting the available niches in which insects can live, and destroying gravel beds where trout can spawn.

The volume and consistency of water flow has a direct effect on trout fishing. Often the maximum life a stream will support is dictated by its minimum annual flow. A stream might be strong and vital for eleven months of the year and then suffer tepid low flows for a single month. All the life within the stream will suffer, and its total trout population will be limited to the number of trout it can support in the short low-water period. The limiting factor to life when water is low, which is usually during the

The structure of any stream dictates where its trout will hold, which will be where their three primary needs are met. In this case, the stream has drawn its own line of greatest depth and most abundant food delivery right down the length of the pool, demonstrating exactly where trout might lie. They would clearly be in association with that curved current seam. The smallest trout would be toward the tail end of it, where both food and depth peter out. The largest would be tucked into that bend at the upper end or hanging just below that rounded point. That is where the great Trevor Hughes has delivered his dry fly and where he took from this pool, on the upper Boulder River, the largest trout . . . not a monster, but nice for the size stream from which it arose.

hottest season of the year, is that killing combination of high temperatures and low oxygen levels.

Maximum flow can also be a problem on some streams. In watersheds that are logged or overgrazed, a storm and its consequent unchecked runoff can be fatal to life within the stream. Gravel and cobble shift from one riffle to the next. Insect life gets ground up. Many trout are unable to find protection from the current and do not survive. Some trout will always survive, but in damaged watersheds the total number will not be what it was before the watershed was damaged, when the stream was protected from sudden high flows.

Artificial interruptions to stream flow can be just as damaging as either high or low flows. Dams are often the cause of both and can have several harmful effects. They halt migrations of spawning salmon, steelhead, and trout, a factor that has done considerable damage to fishing in the Northwest, where many of the greatest runs of fish have been destroyed forever by dams they could not pass to reach the spawning beds above. Dams also turn the inundated stream mileage into lake.

But dams can sometimes be beneficial. The many great tailwater fisheries are evidence of this. The San Juan in New Mexico, the Bighorn in Montana, and the Green River in Utah are all prime examples. These offer some of the finest trout fishing in the world today, and the best of them arise from waters that were historically warm-water fisheries before the dams were built.

Dams that create tailwater trout fisheries release colder water from the reservoir, reducing midsummer temperatures by several degrees. They act as silt traps and release clean water to the reaches below. They stabilize flows, reducing brutal spring runoff and increasing midsummer low water to livable levels for trout. Plankton, which does not grow in moving water, thrives in the still water above a dam, and then enriches the river downstream when it flows over the dam. Certain aquatic insects, for the most part net-spinning caddis larvae, are able to capitalize on this new form of feed in the stream, and their populations explode below dams, which can make trout fat and content.

The availability and activity of trout food always have a lot to do with where you will find trout and what they will be busy doing when you find them. A bit of knowledge about trout foods is therefore important in your ability to read trout water.

3

Trout Foods

The need to feed often overrides the other needs of trout. They will expose themselves to the force of the current, the danger of predation, sometimes even uncomfortable temperature and oxygen levels in order to eat.

Trout move off their stations to feed and even out of their territories if a heavy hatch occurs in water where they would not normally hold. This is especially true on meadow stream and tailwater flats, where the water is thin, overhead protection is scant, and hatches tend to be prolific. If enough insects get active and become available on such water, which is a daily occurrence during the season on many rich spring creeks and tailwaters, trout routinely move out to feed on them at substantial risk of predation and pressure from anglers. Of course, they're wary when they do it.

A typical trout territory offers a compromise of the primary needs: shelter from the current, protection from predators, and food. The station within the territory is a specific spot in which the trout is rested and protected, yet from which it can survey its territory to dash out for the tidbits a constant current delivers. When aquatic insects or other innocents become active, and therefore begin to be available in greater numbers, trout get excited. They leave their stations to dart here and there throughout their territories. If their victims are concentrated at one level, trout might suspend themselves at that level and remain where they can feed with the least movement, the least effort, the least cost in that energy formula. If enough food becomes available outside their territories, they'll move to new stations where they can take advantage of it. When that availability ends, they'll return to their normal territories and their stations within those territories.

Trout are most active when the organisms on which they feed are active as well. Any feeding fish is more receptive than a snoozing fish to an artificial fly, especially if the fly has some resemblance to the food form on which the trout is already feeding. Because trout feed most often on specific insects, baitfish, and crustaceans, a basic understanding of those foods is helpful if you hope to locate trout and catch them consistently.

Insects Help You Locate Trout

The aquatic insect hatches important to trout, and thereby trout fishermen, are covered thoroughly in the companion book to this one: *Handbook of Hatches*. The intention in that book is to help you recognize the insects, understand their behavior, choose fly patterns to imitate them, and present those flies to feeding trout in such a way that they fool the fish, which allows you to catch them. The purpose of the book in your hands is to help you read water and find trout at times when they're not rising or holding where you can spot them. In that case, you locate them with visible clues, usually from the shape of the water. But never neglect to include aquatic insects among the bits of evidence about where trout might be found.

As an example, stoneflies, including the salmon fly nymph and adult pictured here, migrate to shore as nymphs and crawl out of the water before emerging as adults. They do this en masse. Trout follow the migration to feed on nymphs gathered along the shoreline and later on clumsy adults that plummet to the water from streamside vegetation. Your only visible clue might be the presence of stonefly adults in brush and grasses along the banks, but your knowledge of insect behavior will inform you that there has been a recent nymph migration, trout have followed it, and they're likely concentrated right along the edges in good bank water.

Further knowledge of both insect and trout behavior would inform you that trout, for some reason that they keep to themselves, seem to prefer eating stonefly nymphs as opposed to adults. The presence of a few adults, or even a modest abundance, along a stream or river tells you the hatch has started. But it also tells you that many nymphs are still down there, hiding under rocks, waiting for the right moment to

I wrote about all the aquatic insects important to anglers in *Handbook of Hatches*, the companion book to this one. That book tells you how to identify each stage of the various aquatic insect orders, to a level that is useful in angling. It also details fly patterns to match the naturals and suggests the best presentations to show those imitations to the trout as if the flies were the real thing. It provides a thorough treatment of a subject that is vital, and which I will introduce in this chapter as it relates to reading water and finding trout.

Aquatic insects spend most of their life cycles in the water as nymphs or larvae, the stage in which they do their eating and growing. This immature stage typically takes up about eleven months of an aquatic insect's

emerge, usually just before dark or even after it. This knowledge—that the presence of a few adults predicts the similar presence of lots of nymphs, which you can confirm by turning over a few stones in the currents that sweep gently along the shoreline—tells you that trout might be willing to take an imitation of the nymph stage of the adult you see hanging around in the vegetables, even though they'll refuse a floating imitation of the adult.

At some point as a major stonefly hatch moves through its cycle, trout turn their attention upward and focus on adults rather than nymphs. This happens as the number of nymphs dwindles while the abundance of adults increases. If the stoneflies are large ones, such as salmon flies or golden stones, you'll usually hear the transition in the form of rises along the edges that sound like detonations. You'll have no trouble deciding when to switch to dry flies.

The presence of insects in streamside vegetation not only has helped you locate the trout, but also, in this case, has informed you which stage of them to imitate during the unfolding of the hatch.

one-year life span. But there are exceptions: Some aquatic insects, mostly small midges and mayflies, mature and emerge within a few weeks. Others, mainly a few large species of stoneflies and caddisflies, remain in the aquatic stage for two, three, or even four years.

At maturity, the aquatic insect emerges from the water into its winged adult form. This is the stage in which it does its mating, egg laying, and dying. This phase of life lasts anywhere from a few hours to a few weeks. The typical adult aquatic insect has only one thing on its mind—reproduction—and does very little feeding. Mayfly adults do not feed at all; they lack mouthparts and digestive tracts to take in and process nourishment.

Mayflies historically have been considered the most important aquatic insects to anglers. Part of this is because they are the most noticeable insects, when they hatch, and are in many eyes the prettiest of the aquatic insects. But a much larger part is due to their concentrated hatches, which cause trout to feed with abandon and with such selectivity that they notice nothing else, including your fly if it does not resemble the mayfly that is hatching. If you get into a heavy mayfly hatch, it's often easy to read the water and find the trout, because you can see their rises, but it's not often easy to catch those trout if you lack an imitation of the mayfly that is causing the rises.

Mayfly nymphs have adapted to different types of water. Slender and darting *swimmers* live in the greatest abundance in slow water. Though some are found in riffles, the most astonishing numbers thrive on spring creek and tailwater flats, where currents are peaceful and aquatic vegetation takes root. Swimmer mayfly nymphs include the widespread and important blue-winged olives, or BWOs, in the small but prolific *Baetis* complex of mayflies. They browse and flit around on the stems and leaves of submerged plants. They're usually size 16 to 20. Some very large swimmers, such as the size 10 or 12 gray drake nymphs (*Siphlonurus*), tuck themselves back beneath undercut banks, where they swim like little minnows and where some large trout also take roost and are happy to feed on them.

Robust *crawler* mayfly nymphs prefer faster water. They are found most often in riffles and runs, clambering among the cobble and stones of the bottom, hiding in the cracks and crevices that form between the stones, sometimes crawling deep into the streambed. They have weightlifters' strong arms with which to hold on in the current while they browse the thin layer of photosynthetic growth on bottom rocks. A few crawlers, such as the western green drake (*Drunella*) and pale morning dun (*Ephemerella*), have adapted to life in spring creek vegetation.

Flat *clinger* mayfly nymphs use the shape of their bodies and the nature of hydraulics to live in the fastest riffles, rapids, and cascades. Like crawlers, they eat the same layer of algae that we slip on and curse while we wade streams. Their bodies are shaped so that they are able to live in the very thin slice of water that is slowed by friction as it passes over the surface of a streambed stone. Many of these, including the western march brown (*Rhithrogena*), migrate to calmer water before emerging as duns. Others, such as the pale evening duns (*Heptagenia*), move to edge waters and emerge from the nymphal exoskeleton on the bottom in water inches deep, completing the swim to the surface as duns. Trout follow these clinger nymphs and feed on the emerging duns with greed when they can get them.

Burrower mayfly nymphs live in the slowest moving water, in and on sand, silt, clay, or firm mud bottoms. Some burrower species, such as the eastern green drake nymph (*Ephemera guttulata*), wriggle into sand or pebble substrates until only their eyes show. Most burrowers, including the widespread Hex (*Hexagenia limbata*), dig U-shaped tunnels and live in them beneath the bottom, coming out to feed across the bottom surface at night. These require bottoms of a consistency soft enough that they can dig a burrow into it, yet firm enough that the burrow will not collapse while the insect is inside it.

All four of these mayfly nymph types—swimmers, crawlers, clingers, and burrowers—live in slightly to vastly different water types. Each type causes trout to become active on their stations or move out of their territories and onto feeding lies when an abundance of them becomes available.

When mature, mayfly nymphs swim or float to the surface, where the nymphal exoskeleton, or skin, splits along the back and the *dun* emerges. Duns have upright wings, long slender bodies, and two or three long tails. They look like brave little sailboats adrift on water too big for them. As soon as their wings are dry, duns leave the water and fly to nearby vegetation, if they are not eaten by trout first or taken by birds along the way. Mayflies tend to emerge in great numbers on specific water types, and trout do their heaviest feeding when a hatch, the transition from nymph to dun, is happening. The way you use mayfly duns to read water is by looking for rises whenever you notice an abundance of the insects.

There are three moments of vulnerability to the mayfly during a hatch: first, as the nymph makes its way to the surface; second, as it emerges through the surface film and the dun escapes the nymphal skin; and third, as the dun rides the current, waiting for its wings to dry. The individual insect is largely helpless, almost entirely at the whim of the current, the trout, and then the birds. The survival of a mayfly species is dependent on the sheer mass of individuals that emerges at one time, overwhelming predators. If enough emerge at once, at least a few are bound to survive, even in the most parsimonious times. This sort of survival tactic, mass emergence, is an obvious benefit to hungry trout, and trout seem always to be hungry, even when they're already stuffed.

A mayfly dun that survives predation and escapes to streamside vegetation casts a final thin skin and turns into a *spinner*, the reproductive stage of the insect. Male spinners form swarms, often dancing in clouds above streams in morning or evening. When a female mayfly spinner enters the cloud, she is quickly coupled. Her eggs are fertilized in the air, then deposited onto the water. Female mayflies generally die spent on the water in what is called a *spinner fall*. Like duns, spinners are very often available to trout in great numbers over a short time span on a limited reach of water. These falls usually, though not always, happen in early morning or late evening. Male spinners might light on the water with the females, but most return to vegetation and die there, of no value to trout.

Trout feed heavily and selectively during a mayfly spinner fall. But they know that the insects are trapped on the water and are not about to escape, so they often rise to sip them gently. Such feeding can easily go unnoticed if you're not observant, especially in low light, when many spinner falls take place. If you see mayfly spinners in the air, read the water by watching it very carefully for the tiniest rises, which can conceal the largest trout.

Caddisflies live most of their typical one-year life cycle in the water as wormlike larvae. These larvae take one of two forms. Many are *cased*, carrying around homes built of sand and sticks and stones. Cased caddis larvae generally live in the slower water of runs, gentle vegetated flats, and pools, though some build their cases of heavy material for ballast and live on the bottom in fairly fast currents. Large fall caddis larvae (Limnephilidae) are examples of these cased caddis types that can live in brisk flows.

Other caddis larvae are *free-living*, not building portable cases. Some free-living caddis larvae construct crude shelters along the bottom or in rooted vegetation, into which they can retreat, and spin nets that filter particles out of the current, on which they feed. These *net spinners* (Hydropsychidae and others) are abundant in riffles, runs, and flats with sufficient current to deliver particles of food. Other free-living caddis larvae roam the bottom without building any shelter at all. These are predaceous, feeding on anything that they can subdue and eat, usually midge and blackfly larvae or mayfly nymphs. These free-living caddis larvae are most abundant in the fast water of riffles and brisk runs. They are almost always some shade of green, hence their common name green rock worms (Rhyacophilidae).

Caddis pass through a transitional pupal stage before becoming adults, in exactly the same manner that a caterpillar spins a cocoon and passes through the chrysalis stage before becoming a butterfly. When ready to pupate, cased caddis larvae attach their cases to bottom stones, vegetation, or woody debris, seal them, and spin cocoons of silk inside. Net-spinning larvae seal their shelters and spin a cocoon inside the retreat. Free-living green rock worms construct rough shelters of pebbles and sand grains and then spin their cocoons in safety inside.

Pupation, or the metamorphic change from larva to adult caddis, takes place in the pupal cocoon. The larval stage of the life cycle lasts most of a year; the pupal stage lasts a few weeks. When this transformation is complete, the fully formed adult caddis is encased in the pupal skin, still inside the cocoon. When ready for emergence, the insect exudes bubbles of nitrogen between the inner adult skin and the outer pupal exoskeleton, helping separate the skins so the adult can escape more swiftly when it reaches the surface. These bubbles also help the pupa—or *pharate adult*, as it is more correctly called in this brief transition—rise quickly to the surface.

The insect uses scissorlike mandibles to cut its way out of the inner cocoon and outside shelter or case and then either drifts along the bottom for some distance or begins rising toward the surface at once. In some species, this gas-assisted rise to the top is aided by a strong swimming motion. Other species merely ascend at the whim of the currents. In either case, they are very vulnerable to feeding trout. It's likely that as many caddis are taken during this brief trip from the bottom to the top, as pupae, as are taken as larvae in the long period leading up to the pupal stage, or as adults going about the briefer business of mating and sowing the seeds for the next caddis generation.

Caddis pupae break through the surface quickly and cast their pupal skins, and the adults fly off toward streamside vegetation, most of them in a hurry. Though a few caddis types, such as the black caddis and Mother's Day caddis (Brachycentridae), might stay on the water for drifts of five to twenty-five feet, most get off the water more quickly than the average mayfly dun. These escaping adults leave a lot of splashy rises in their wakes. You need to read these rises to locate your trout, but you also need to know that most of those vigorous rises are to pupae, or pharate adults, that are just about to escape through the surface, rather than to adult caddis taken off the surface itself. As you're fishing, you are seeing rises and adult caddis in the air, but a floating imitation will often be refused. What you need to fish is a soft-hackled wet fly or a caddis pupa imitation.

The moment of most vulnerability for an emerging caddis is the brief pupal trip from the bottom to the top. Danger from trout is especially concentrated in the instant before the insect reaches the surface to break through. When a caddis hatch of any magnitude is on, trout often leave their stations and hang suspended a foot or two beneath the surface, waiting to intercept helpless pupae. If you see caddis adults in the air, but no signs of surface-feeding trout, you should guess the presence of pupae at a very minor depth and once again drift a pupa pattern or swing a soft-hackled wet fly.

Caddisfly adults have tentlike wings pitched over fat, bulbous bodies. They mate in grasses, shrubs, and trees alongside creeks, streams, and rivers. Many species, such as the large fall caddis, return to the water to deposit their eggs by dancing and tapping their abdomens to the surface, washing off the eggs in clusters. Many trout stream caddis, such as the most common net spinners *Hydropsyche*, or spotted sedges, and *Cheumatopsyche*, or little sister sedges, dive through the surface of the water, swim to the bottom, and deposit their eggs on bottom stones. Caddis that deposit their eggs on the surface cause excellent dry-fly fishing. Those that deposit their eggs on the bottom, once you understand that trout might be splashing at the surface but are most often taking swimming adults beneath it, can give you excellent wet-fly fishing. Again, you read the presence of caddis adults to mean trout feeding a foot or so deep.

Stoneflies have a simple life cycle, undergoing incomplete metamorphosis, with no pupal stage. They live in the water as nymphs and then crawl out of the water before emerging from the nymphal skins as adults. Most stoneflies have a one-year life cycle, but a few of the largest types, golden stoneflies and salmon flies, have three- to four-year cycles.

Stonefly nymphs thrive in water that is well oxygenated but do very poorly in water that becomes warm and lacks oxygen. They are found mostly in riffles and runs; their preferred habitat is bottom cobble and rocks washed with strong currents. Many stonefly types are peaceful herbivores, browsing the thin layer of algae that coats almost all rocks in running water. The largest of them, the salmon flies (Pteronarcyidae), are among this group and are sometimes referred to as the buffalo herds of streams. Other types, such as the golden stones (Perlidae), are predaceous, prowling through the crevices among bottom stones, hunting for mayfly nymphs, caddisfly larvae, and any other organisms they are able to capture, kill, and eat.

All stonefly nymphs, when ready to emerge as adults, migrate along the bottom to shore or protruding boulders and crawl out of the water as nymphs. Their emergence does not take place out in open water, as it does in most mayfly and caddis species. Emergence takes place away from the water, safe from predation by trout. This often happens at night, safe from bird predation as well. Some stonefly varieties emerge at the waterline, very near the water. Others crawl several feet from the water as nymphs, before getting a grip on something and leaving their nymphal shucks behind. A few, notably the golden stones, might crawl fifty feet from the water and then ten to twenty feet up a tree before emergence.

Stonefly adults of almost all types hang around in streamside vegetation for a week or more before mating and returning to the water to deposit their eggs. During this period, it's common for them to fall to the water, right at the edge. Trout, already drawn to shore by the migrating nymphs, remain by the edge, waiting for adults to fall into the bank water. It's interesting to note that trout will continue to focus on nymphs of the larger stonefly types, especially the salmon flies and golden stones, as long as they're available, ignoring adults, sometimes even those that fall to the water right over their heads.

Both the migration of stonefly nymphs and the presence of stonefly adults at the edges draw trout to bank water. If you notice even a few adults in the vegetation, suspect that trout have moved to follow them. Try dry-fly imitations first, as we all do, but switch quickly to nymphs fished within a foot or two of the bank if your dry flies fail to entice trout.

What most people think of as a stonefly "hatch" is actually the afternoon or evening return of females to lay their eggs. They hover over riffles or broad runs, helicoptering in to wash egg clusters into the water. This can be fatal behavior, especially when the insects are among the largest of stoneflies and therefore the most exciting to trout. If stoneflies are in the air and you see or hear detonations in the water, you know that trout are taking adults and will have no trouble reading the water—simply follow the sight or sound of the rises and rig appropriately to imitate them.

Midges are likely important in the diet of all trout, especially fry and fingerlings, in all types of water. But they become important to anglers on moving water most often where currents are less than brisk and bottoms are more silted than they are freestone. Midge larvae are most abundant where they can burrow at least slightly into the bottom. That means you'll find midges most important on spring creeks and tailwaters, least important on mountain and foothill streams. In the valley reaches of mature rivers, where the water has flattened out and lost most of its urgency, midges can be abundant in long pools, flats, and runs. You'll rarely fish a midge imitation in a riffle.

It's important to note that while some still-water midges become big enough to be called "bombers," midges in moving water tend to be small to tiny. A big one, in my experience, is a size 16. Much more common are midges in sizes 18 and 20. Many of the midges I fish over on tailwaters are imitated with flies tied on size 22 and 24 hooks.

The midge life cycle is similar to that of the caddis: They live as larvae, swim and are buoyed to the surface by gases as pupae, emerge and deposit their eggs as adults. The wormlike larvae are not of great interest to anglers, as they're generally tiny and live inside a burrow in the bottom where trout can't get at them. Larvae are not available to trout in the concentrated numbers that cause trout to feed selectively on them.

Midge pupae are extremely vulnerable. A very high percentage of midges eaten by trout are taken in this stage. Pupae rise feebly to the surface and then hang suspended and helpless beneath the surface film. They might be tiny, but so many can be available during a hatch that trout set up feeding stations and gobble them eagerly. The pupa is the most important stage in the midge life cycle. Trout often feed selectively on midge pupae but rarely on midge larvae or adults. You read water during a midge emergence by looking for rises, but remember that trout feed most often on pupae and those rises are just inches beneath the surface rather than in it.

On rare occasions, midge adults remain on the water and gather into clusters. This almost always happens on the edges of winter, or even during winter, when the fragile insects have trouble getting airborne. They accumulate into rafts, and trout seem to key on these gatherings, neglecting individuals. At such times, you can have excellent fishing over rising trout with cluster midge patterns, and the presence of the midges is extremely helpful in reading the water, because the rises let you locate trout instantly. I've had cluster midge fishing just a few times, most notably on the Bear Trap Canyon section of the lower Madison River in an April snowstorm and on the Cimarron River in New Mexico during a February rain- and windstorm. The lower Madison is a mature river stabilized by lakes, and the Cimarron is a tailwater.

The best midge fishing I've had on moving water has been on the Bighorn River in Montana, where the insects I've imitated are size 22, and on the Deschutes just below the dams, where they are much smaller. Both rivers are tailwaters. In both cases, trout were feeding selectively on pupae, not adults. Neither situation seemed to require a storm for good fishing to happen. In all cases where I've had good midge fishing in creeks, streams, and rivers, the midges have helped me locate trout only by their rises.

Other aquatic insects of occasional importance in running water include alderflies, hellgrammites and closely related fish flies, and water boatmen. Alderflies are caddis look-alikes that are more important on ponds and lakes than they are on streams. Hellgrammites and fish flies are the larval stages of what will later become dobsonflies and adult fish flies. They are very large and fierce; they will be happy to bite you if you give them the chance. They are most important on midwestern and eastern trout streams. I've collected them in greatest numbers in beautiful North Carolina mountain streams. But they do not help you locate trout except by predicting, through their presence in a collection sample, that trout will be on the bottom and willing to take a tumbled imitation if nothing is going on to draw the fish's attention in a different direction.

Water boatmen are largely still-water insects, but they do live in slow, weedy water in streams, and trout occasionally feed on them in such water. If you're fishing a spring creek or tailwater with beds of rooted vegetation, be on the watch for trout nosing into the plants, prodding out water boatmen, and feeding on them as they try to escape. In terms of reading water, water boatmen help you by causing the trout to expose themselves, often with the subtlest wink of a flashing flank as one turns to kill a stroking boatman.

Terrestrial insects can be of great importance to the trout stream angler. Grasshoppers often find their way to the water. They get into most of their trouble with trout in hot weather during July, August, and September on meadow streams. Grassy bankside vegetation that grows close to the stream favors the chances that hoppers will get delivered to the water and thereby to trout. Undercut banks and good holding water along the edges promote the chance that a trout will be holding right there, waiting for a fallen grasshopper. Forested streams are far less likely to be important hopper water. The presence of good numbers of grasshoppers in conjunction with the type of water that is deep near shore, providing protection from predators, is your prediction that trout will be found tucked up against those edges.

Beetles, on the other hand, can be most important where a forested shore increases the chance that they will be blown out of trees in great numbers and wafted to the water. Other important terrestrials include ants and crickets, leafhoppers and inchworms. The presence of these terrestrials, again, predicts that trout will be holding along edges that give them protection from overhead predation.

Almost any insect, or even small land animal such as a mouse or vole, that gets onto the water is fair game for fish if the victim is small enough and the trout large enough. Jim Schollmeyer and I once fished the Blackfoot River in early summer. He walked a gravel bar and saw a baby killdeer take to the water ahead of him, ignoring the frantic cries of its mother to return to shore, and disappear in a monstrous boil. Jim couldn't bring that trout back to any fly he tried. It was likely far down in its bomb shelter, digesting its big meal.

Baitfish, sculpins, and young-of-the-year trout are important food items for trout. The larger a trout gets, the more attention it turns to these bigger bites, and the more it seeks the kind of territory that provides them. In terms of reading water, the presence of these prey fish tells you that big brown trout might be out hunting on tailouts of pools and in the shallows along shore right as nightfall nears, places where they would not risk exposing themselves in the light of day. But you won't find such trout, or such activity, in water where no bomb shelter is near.

Crustaceans are also important trout foods. Scuds are small shrimplike beasts that pedal furiously, upside down, through the weed-filled reaches of our slower rivers. These comical creatures are abundant in favorable habitats and available all year round, and trout search for them hungrily whenever aquatic insects enjoy a day or a season off. Aquatic sow bugs, or cress bugs, are important on a few limestone spring creeks. They are most abundant where the water is rich in calcium, the flow choked with plants.

The presence of either scuds or sow bugs, like water boatmen, should encourage you to watch for trout feeding among rooted plant beds.

Crayfish are an obvious big bite for trout that live in large pools. Studies of stomach contents reveal that crayfish make up a substantial percentage of trout diets, and that the larger the trout, the higher the percentage of crayfish it eats. As with baitfish, the presence of crayfish should prompt you to look for trout in tailouts and edge shallows, but only at dawn or dusk.

Leeches are largely still-water inhabitants, but they live in the slower stretches of streams and even worm their way through the gravel of riffles. They make up a very small percentage of the feed in a trout's diet in moving water, but fish seem to have a long memory for them. You can draw a wallop out of a large trout with any fly, such as a Black or Olive Woolly Bugger or Bunny Leech, that imitates their sinuous snakelike swimming motion. It's difficult to say that the trout takes the fly for a leech, but it's not difficult to say that such an imitation fished on mature rivers, especially along slow bank water or in the depths of runs and pools, will draw some whacks from very large trout.

In his Scientific Anglers video *Anatomy of a Trout Stream*, aquatic entomologist Rick Hafele, author of *Nymph Fishing Rivers and Streams*, emphasized that "identifying the insect trout are feeding on, and its life stage, will determine at what water level you fish." Learning a little about all the trout foods will tell you what sort of imitation to choose and what tactic to present it with, as well as where to fish it in the stream.

Trout foods help you learn to read water in two major ways. The first is during a hatch, when feeding trout are exposed by their rises and you can find them, though you need to watch the water carefully. The second is by the presence of a specific insect, which will often predict that you will find trout feeding in specific areas, such as along the banks when stonefly nymphs are migrating to shore for emergence or deep in plant beds when an abundance of water boatmen, scuds, or aquatic sow bugs become available to marauding trout.

The behavior of trout foods is one of the elemental keys to the behavior of trout. Learn what you can about them, observe as much as possible about the relationship between trout and trout foods while you're out on the water, and you will increase your ability to read water and find trout.

Trout Lies

Trout hold in four types of water: *sheltering lies*, *holding lies*, *feeding lies*, and *prime lies*. Each offers something different to trout, and each satisfies trout in a different way. Learning to read water and recognize the types of lies it provides will improve your ability to locate trout.

SHELTERING LIES

Sheltering lies are sanctuary water. They are what Charlie Brooks referred to as bomb shelters. When you approach a pool carelessly and see trout scattering before you, they are arrowing off to these lies.

We normally think of a sheltering lie as the deepest part of a big pool, where fish cover their eyes with their fins and hide down in the darkness. But in a mountain creek, a trout might dive under a log, tuck under a boulder or ledge, or find a fragment of shade and lie in it without motion. In a meadow stream, a trout's sanctuary might be far back beneath an undercut, among tangled roots. In a spring creek, the same trout might merely sink down to disappear among trailing plants. On big rivers, trout can usually find sanctuary just by moving a few feet out from the bank and dropping down to the bottom.

If trout have fled in fear into their sanctuary water, it is little use fishing for them without giving them some time, ten minutes to half an hour or so, to calm down, forget what sent them there. But if you approach a recognizable bomb shelter and it's undisturbed, you should consider it a holding lie and fish it carefully. The same water that provides a safe haven for panicked trout of normal size might serve as a permanent home for a single large one. After all, the old lunk will often get its groceries driven right to it by careless waders.

Though we tend to think of a typical sheltering lie as a deep pool, it's just as likely to be a logjam lodged across the current, such as this one on Oregon's Metolius River. True pools are not as common in streams and rivers as we commonly assume; they're most abundant on small mountain waters, where plunges dig pools. But some sort of shelter—logs, a trench, the root wad of a fallen tree—is normally near in most trout water, and trout will dive into it when threatened by an otter or osprey, or by you and me if our wading and presentations are clumsy enough to drive them there.

Trout holding in the convenient currents downstream from this jam, crouched on the bottom in the riffle outside of it or feeding on the flat upstream from it, will find plenty of places to dive into complete protection underneath it if they feel threatened. The nearness of such a shelter will often cause holding or feeding lies nearby to hold more trout, and larger ones, than lies that lack some local sanctuary into which they can flee.

Because reading trout water is as much a matter of intuition built on experience as it is book learning—see page 2—I've added silhouettes of trout in all the photos, based on where I've either found trout, or would expect to find them, in the water shown. If you spend a moment focused on each trout in each photo, your brain will add that situation to your memory bank and will bring it back to you in a blink next time you're out on a creek, stream, or river, reading its prospective lies to locate trout.

This pool fished by Jay Nichols, on Cedar Run in Pennsylvania, could be called a typical bomb shelter. Any trout feeding along the edges of the pool, or in the run downstream, would head into the depths beneath that whitewater at the first sign of danger. But the same piece of water would deliver food from upstream, provide shelter from constant currents because of turbulence over and around rocks along the bottom, and offer protection from predators because of the disturbance of the water overhead. It could therefore be a prime lie for what might be the largest trout in this reach of the stream.

Water that forms a sanctuary for some trout can be considered a prime lie for others. Consider that Cedar Run holds a few big brown trout that are mostly nocturnal, think about where one of them might lie in wait for evening on this bit of stream, and you might begin to wonder which predator a ten-inch trout would be smarter to avoid. It's likely a lazy brown trout would tuck in under the ledge at left, where it would find plenty of depth to hide it and little current to fight.

HOLDING LIES

Holding lies are where trout are found most often. Such lies offer at least some shelter from currents, more than modest protection from predators, and a fair crack at feed delivered on the currents—the kind of compromise that makes up most trout territories. These are the kinds of lies in which trout spend most of their time. A trout establishes its territory, waits on its

Most holding lies are marked by some sort of obstruction to a current that trout otherwise would not be able to fight continually. Such lies are not always obvious. But most holding lies send some sign of themselves up to the surface, which always reflects what's on the bottom. In this case, the water on Colorado's Big Thompson River, fished by *Fly Fisherman Japan* magazine's editor Terri Yamagishi, shows obvious signs of a line of boulders, while the water in which he is wading lacks evidence that the bottom currents might be broken.

station, and feeds on whatever the current supplies it. Most holding lies offer less-than-perfect protection for the fish they hold and provide less than all of the food they need. Studies have shown that shelter from currents, not food, is most often the limiting factor in the suitability of a territory. The better the shelter, the better the territory. The value of a holding lie is also increased by the nearness of a sheltering lie, where a trout can retreat in a hurry when its holding lie gets invaded by an osprey, otter, or angler.

Trout, as they grow larger, move through a succession of holding territories. When a trout outgrows one territory, it will dislodge a smaller or less aggressive trout from a better territory, and another trout will immediately take over its abandoned territory. Reshuffling is probably constant as trout grow older and either move up to a larger territory or die wishing they could, are driven out of territories by other trout, or are captured by predators, which might include us.

Many stretches of unfeatured water, such as this run on the Yellowstone River fished by Jim Schollmeyer, have the potential to hold trout anywhere along the bottom. Boulders that interrupt the current are too deep to send signals of their presence up to the top, but you know they're down there because that surface is so choppy, and you know trout are down there too. You just have no idea precisely where they might be, so you must set up a disciplined casting pattern that shows your fly or flies—whether a dry, nymph, brace of one or the other, or combination dry fly and nymph dropper—to all the water.

A holding lie almost always has some physical feature that separates it from the surrounding water. It might be a rock or boulder that breaks the current, a seam where two currents meet to deliver more food, a trench where the water is a bit deeper than that all around it, or an undercut bank where the current is slowed but a trout tucked back there can dash out to take whatever the current contains. Any feature that interrupts the current, in water that is deep enough to give even a slight sense of security, will attract a trout of one size or another. In general, the more the current is interrupted and the deeper the water there, the larger the trout a holding lie will be able to brag about.

Holding lies always have something to distinguish them, from the trout's point of view beneath the water. But those distinguishing features are not always visible from our point of view above the water. Not all rocks and seams and trenches in a three-foot-deep riffle are reflected so that they show on its surface. Some boulders that stud the bottom of a deeper run send boils up to the surface; most do not. A boulder the size of a basketball can make a pretty good lie for a sizable trout. You might notice signs of that boulder in a shallow riffle but likely will never know it's there in a deeper riffle or run.

Sometimes you must learn to recognize trout *water*, as opposed to trout *lies*, and fish all of the water as if it all were a lie. Trout water is distinguished by a nature that promises lots of hidden holding lies, but it doesn't necessarily reveal any of them. Unfeatured riffles, two to three feet deep, often fall into this category, with trout scattered across them in territorial niches that don't show on the choppy surface. Some runs that are three to six feet deep and unbroken on the surface hold lots of trout without exposing the exact location of any single one of them. By setting up a searching pattern of casts that covers all these kinds of water, you will deliver your fly to any trout that are holding constant on their invisible stations, waiting for bits of feed to come to them or to float over their heads.

FEEDING LIES

Feeding lies are places where trout leave their stations, and even their territories, to go to the food when it's abundant, rather than wait in their holding lies for food to come to them. Feeding lies offer an opportunity for trout to get greedy and let them more than break even in the energy exchange formula, if it's necessary for the fish to fight a current. But such lies usually offer little protection from predators. If they did provide protection from predators and shelter from currents, trout would hold in them all the time, in territories and on stations, and they would be holding lies or even prime lies, not feeding lies.

Feeding lies are usually areas of the stream that are rich in insects, sending up prolific hatches. This kind of water is almost always shallow, because the greatest amount of sunlight strikes the bottom where the water is shallowest, resulting in increased photosynthetic plant growth, and that plant growth feeds aquatic insects, which feed trout. Sometimes the water is deeper but has rooted vegetation and smooth surface currents. When trout move out of their normal holding lies to feed in shallow water, or when they set up stations high in smooth water so they can sip insects from the surface with less expenditure of energy, they are vulnerable to overhead predation. They know it and are wary. But if the hatch is heavy, trout sometimes concentrate so intently on feeding that they seem to forget

Exposed Feeding Lies

This water, on the Bighorn River in Montana, is too shallow and smooth to offer trout protection from overhead predation. But it's a tailwater, with stabilized flows, rooted vegetation, billions of microniches for aquatic insects and crustaceans, and sometimes incredible hatches when they happen. When trout are not feeding, they move out into deeper water and drop down into channels in the weeds or holding lies behind the few boulders in the current. But they cast all fear about osprey aside when insects hatch; then they move in pods onto the sort of shallows where Jim Schollmeyer has his rod bent. When they do, such water becomes a feeding lie. You read it by looking for rises and collect what the trout are taking before selecting a fly pattern or settling on a rigging that might be wrong if you don't do some slight study.

The black caddis (*Brachycentrus americanus*) is an exceptional hatch on the Bighorn River in fall. A few minutes spent collecting will provide dozens of samples to study. Trout move onto feeding lies to work on them in pods of ten to sometimes twenty fish, one pod all rainbows, another all browns. When they do, you read the water by pinpointing rising trout. Then you match the insects and focus on refining your tackle and presentations.

about raptors in the air or anglers in the water. During a hatch of mayflies, midges, or caddis, it's common to have trout go right on feeding and ignore you while you wade very near and cast to them, especially on heavily fished waters.

PRIME LIES

Prime lies are places along the course of a stream where all the needs of a trout are met in one place. These lies offer comfortable temperature and oxygen levels, easy living sheltered from strong currents, protection from predators, and an abundant source of food without the need to move away from the established territory. A trout holding in a prime lie does not ever need to move far to satisfy any of its needs.

The markings of a prime lie are usually, though far from always, visible on the surface of the stream. It generally has a steady current that delivers food, though in a pool it might be a ponderous current. There is always some significant break in the current, though you might not always be able to see it. This interruption to the current can take the form of a protruding or submerged boulder in a riffle or run, a ledge across the head of a riffle, a trench where the bottom drops away in brisk water, or anything else that slows the current so a trout can hold comfortably. The final sign marking a prime lie is either sufficient depth to give security from predation or a rough surface that breaks up the sight line of kingfishers, in the case of a small stream with small trout, or ospreys, in the case of larger water holding larger trout.

Recognizing prime lies takes some practice, but when you keep in mind the three needs of trout they provide—current to deliver food, shelter from that current, and protection from predators—it becomes relatively easy. When you find a place where all three of these needs are met, you'll usually find the largest trout in this section of stream.

Fishing prime lies is not always easy. In fact, a fourth factor that points to a prime lie in heavily fished water is the degree of difficulty you must go through to fish it. The reason is simple: On waters with lots of angling pressure, we become one of the primary predators from which trout seek protection. That is why the person who is willing to struggle across difficult currents and fish the stream from the least common bank will often take the least common trout: a big one. It's also why the angler who squirms into the most difficult indentation along a brushy bank often has the largest tales to tell of trout that got away.

This shelf in the Metolius River, probably the remnants of an ancient lava flow, shapes the water perfectly for a prime lie. The current is broken, so trout would find shelter in the line of current all the way from the slight white tongue at the upper end to the current seam downstream, where Masako is nymphing the water. There is plenty of depth for protection from predators. Food, in the form of aquatic insects emerging on top or terrestrial insects falling from over-head, would be compressed into that current seam and delivered down it.

At times you'd have good dry-fly fishing here, but it would be more productive to fish the bottom in the absence of evidence that trout are rising. Nymphs and larvae of stoneflies, mayflies, caddisflies, and midges would all find niches in the gravel and stones along the bottom. Trout would find this a comfortable place to hold at all times, and they'd be able to venture out to the feeding lies on all the water surrounding it when food forms became more concentrated there.

Reading water is a matter of learning to read the stream's evidence of where trout would be. As Rick Hafele so aptly put it in his video *Anatomy of a Trout Stream*, "Learning to read water means being able to find where the trout's needs are being met by reading the *surface* of the water." You must learn to notice the indications on the surface that spell the different kinds of holding water underneath: sheltering lies, holding lies, feeding lies, and prime lies.

The Various Shapes of Prime Lies

Prime lies come in as many different shapes as do creeks, streams, and rivers; probably many more, because each trout stream has prime lies shaped by each of its different water types. But all prime lies provide for the three primary needs of trout that hold in them: shelter from currents, protection from predators, and food.

In this prime lie, on Pennsylvania's Kettle Creek, food is provided by drift from the riffle upstream, shelter from currents comes from the shelf that cuts across the stream, and protection from predators is furnished by the depth of the pool. If needed, a convenient bomb shelter is a short dash away beneath the roots of that tree, which might indeed be the prime spot.

continued on next page

This photo of Terri Yamagishi, on the Big Thompson River in Colorado, shows an obvious current slot at the edge of the big boulder. The current delivers food. The boulder shoves the current aside. The bottom edge of the boulder forms a protected lie. More than one trout holds here, but the biggest, which Terri hooked and landed, held in the softened current near the upper edge of the boulder and moved out into the faster slot to intercept his dead-drifted nymph. It was a sixteen-inch brown.

Prime Lies continued

The water on the Lochsa River in Idaho is so clear that it opens its prime lies out for easy reading. In this case, the deep slot right under my rod delivers food and offers protection from predators, and all those boulders on the bottom deflect the current. If a trout ever feels threatened, all it has to do is tuck into shade between a couple big rocks on that bouldered bottom, and it would be safe from anything that lacked a jackhammer.

I call this type of prime lie a *corner*. This one, on the Yakima River in Washington, is being probed with a streamer by fly-fishing author Skip Morris. Food is supplied by the vast expanse of riffle upstream. Shelter from currents comes from the shelf that juts into the current and forms the corner. Protection from predators is provided by the depth of the lie and the rumpled surface over the top of it. Skip is demonstrating one of the vital points about fishing such corners with anything on the swing: Get far enough up into the head of it that your fly swings into the prime water. Far too many folks wade into a corner at about the place where most trout hold. Then they cast across current, and their streamer, wet fly, or nymph swims around in water that, although it is good, is not nearly as beneficial as the water upstream, where trout line up to get first whack at whatever the current delivers over that shelf.

EMPTY WATER

A final important kind of water that you should learn to recognize is water empty of trout. This is the largest percentage of water on any creek, trout stream, or river. If you scatter your casts over the water at random or, worse, discipline yourself to cover every bit of water thoroughly, without regard to its potential to hold trout, you'll deploy your flies uselessly more than half of the time, because more than half of the water is devoid of trout. Learn to eliminate time spent fishing empty water and your catch can automatically double. Surprisingly, the math can be exactly that straightforward.

At the start of a five-day float trip in Montana, my brother Gene, new to fly fishing, cast to whatever water was in front of him on his side of the boat as we drifted downstream. Our guide, Don Williams, would call out, "Fish the other side!" from time to time, and Gene would turn around and cast again to whatever water was in front of him.

But Gene's a doctor, and it wasn't long before he began to recognize which bank was best without being told. He would turn his back on flats that were inches deep, to plunk a weighted nymph in tight against a bank where the water was three to four feet deep. At first his casts went at random along the favored bank. Don would coax, "Now just in front of that rock!"

After a while, Gene learned to recognize such prime holding lies and plunked his nymph, on a high percentage of his casts, to features that indicated the few best spots along a bank or to lies marked by boulders out in open water. His percentage of hooked fish turned upward sharply day by day.

But Gene's greatest progress began to occur when he learned to recognize water that was not likely to hold any trout at all. Empty water is almost always featureless. It might be a long, pebbled riffle a foot or less deep, with nothing to interrupt a fast current, or a broad flat that is too thin to protect trout from overhead predators and has no depressions or trenches to provide even minimal shelter where a fish could dive in an emergency. The inside bends of a freestone stream are often flat gravel bars where the water is too shallow to hold trout. Both current and depth are pushed toward the outsides of the curves. Thus most features and holding water will be found against the outside sweep of a bend, and almost all of the trout will be found there too, leaving empty water on the inside of the bend. That's where you should wade to probe the outside edge.

Water that is shallow and featureless is generally empty of trout because it does not offer sufficient protection from predators or shelter from swift currents. But such water is sometimes rich in insect life, offering up concentrated hatches of mayflies, caddisflies, or stoneflies. When a hatch

Learning to read empty water, and the areas where it transitions into water with potential to hold trout, will allow you to cease fishing water where you have no chance and increase your catch merely by increasing the amount of time you spend with your fly in water where trout hold. Note that where Jim Schollmeyer wades this riffle, on the Willamette River in Oregon, he's only ankle-deep. Where he's swinging his wet fly, however, the water levels out, slows down, and deepens to a couple feet. If the bottom is broken enough in the water he's fishing, it will be a holding lie, where trout can be found all the time. If it's not sufficiently cobbled to break the current, then it will form a feeding lie, where trout will move up to take advantage of the abundance of insects that are certain to be drifting daily out of that rich but empty riffle upstream.

The flat farther upstream constitutes one broad and productive feeding lie, where trout move out to feed whenever an insect hatch entices them there. In the absence of feeding fish, obvious by the lack of rises, you should to consider it empty water as well. This location, in my own experience from fishing it often, can be high-graded to the small spot where Jim is fishing unless trout are rising elsewhere, in which case I would park the boat and get out to fish for them. I respond to feeding lies for the same reasons trout do: When insects hatch, trout move to feed on them, and I move to pester the trout.

begins on a stretch of empty water, that piece of water might suddenly become the best feeding lie around. Trout know it and will move into it to feed, even at the risk of a tithe of themselves to ospreys wheeling overhead.

Don't waste time thoroughly fishing what you judge to be empty water, unless you see some sign that trout might be holding in it or feeding there during a hatch. Always examine what you believe is empty water with a careful eye, however, and probe it with a few passing casts. It might not be as empty as you think. Such watchfulness and experimental casts help you fine-tune your definitions of the different types of empty water. They can also provide you some surprises.

While Gene, Don, and I drifted that Montana river, we got whacked by a sudden thunderstorm. We dove for slickers, threw up the hoods, covered our gear with tarps, then drew our necks into our collars as far as we could and sat like turtles waiting for the rain to stop. It took two hours.

When the rain did stop, Gene and I stood up and began casting again. Through a two-hundred-yard reach of river, the water was shallow on both banks, inches deep at the edges and only a foot or so deep about five feet out. There was no better water around, so we both cast idly to these banks, waiting for the boat to slide downstream to water that we considered worth fishing. But brown trout began climbing all over our Woolly Buggers. They erupted out of water so shallow it should not have held them.

Before we had gone the two hundred yards, we had each taken a pair of browns that approached a couple pounds. They had moved up out of deep water with the heavy rain, under protection of the darkening sky, to forage on whatever insects or other beasts might get active along the banks. These trout held in what I normally would have judged to be empty water, especially for trout of their size. But even then, they always rushed out to ambush our streamers from behind some slight feature: a football-size rock, an overhanging tuft of bunchgrass, a cut where two shallow currents came together and disturbed the surface enough to disrupt overhead vision.

Water that is almost still will usually be empty of trout unless it is deep enough to be a prime lie for a large trout. But even in deep water, there must be sufficient current in some part of the still water, usually at the head of the pool, to deliver food. Generally, a pool that is surrounded above and below by fairly fast water will be a prime lie, whereas a long reach of deep but uniformly very slow, or even still, water will not be any kind of lie at all.

There is almost always some line of demarcation between empty water and water that promises to hold trout. It might be a surface seam where shallow and deeper currents meet, a quick deepening of the water that shows as a dark area on the surface, or a boulder or some other break to the current. Some shade or some sweeping overhanging grass from the bank will often be the only feature on a long, flat stretch of water, and it will indicate the only place trout might hold.

Where you notice a demarcation line, or some sort of feature in water that otherwise looks empty, it's always wise to fish it carefully. These are the sorts of signs that denote lies, and lies of all types are where you find trout.

Distinguishing Trout Water from Whitefish Water

Whitefish are common in Rocky Mountain and western states and provinces, less common as you move to the Midwest and East. Biologists report that whitefish seldom compete with trout for the same holding lies or food, though I've occasionally seen them mixed while feeding on a hatch, usually of small mayfly duns. You may find a pod of whitefish feeding in one area of a riffle or run, a scattering of trout feeding nearby. Learning to distinguish trout water from whitefish water is one way to limit your fishing in "empty water," assuming you would rather catch nothing at all than be bothered by whitefish. Wherever you start catching bunches of whitefish, just consider it empty water and jog off to fish somewhere else . . . unless, like me, you would rather cast over rising fish than do almost anything else, and you don't have high standards about what kind of fish might be rising as long as they take flies.

Whitefish prefer moderate-to-fast riffles and runs and seem adapted to surviving in those that are relatively featureless. They won't hold in torrential flows, but they seem not to need obstructions to break the current in water of modest and sometimes even fairly fast velocities, as do trout. It's probable that whitefish hold in the inches-thin layer of water slowed by friction at its interface with bottom rocks. They can be found wherever the bottom is cobbled and populated by an abundance of small life forms.

Whitefish have tiny mouths and will accept small to medium nymphs, size 16 up to 10 or 12, delivered in or just above that narrow zone of slowed water. If the water is shallow, they'll also arrow to the top for small dry flies. They often feed on the surface in the slower areas upstream and down from riffles and in runs during hatches of small insects. They take insects off the surface with a peculiar roll. Such rises appear vigorous and often send a characteristic few drops of water into the air, almost like the pyramid rise of a trout. With practice, you can distinguish whitefish rises from those of trout; on those occasions when they rise close together, this lets you target the trout and minimize the number of whitefish you catch, if that's your desire.

As with all kinds of empty water, it's best not to carelessly dismiss a particular piece of stream as whitefish water without fishing it to see if it is

Reading Cloudy Water

Potential holding lies can be located with fair certainty even when the water is cloudy, as in this case on Oregon's White River, which has its origins in melt from a glacier. When the sun is hot on the mountain, the river turns silty, but its trout don't simply disappear. If you read the water correctly and fish it with diligence, so that trout have every chance to see your nymphs, you'll hook them, sometimes more consistently than you would if the water were so clear that they could see you coming.

Masako took this modest-size trout after probing the opaque run for more than half an hour. Note the large salmon fly nymph dangling in the air, which means the trout took the size 16 trailer, probably a Beadhead Prince, which we fish often on the White River. It's more common to hook trout on the smaller of two nymphs, but it always seems to reduce the catch if we cut back to just that one fly. It's possible the trout see the big nymph, move to it but refuse it, then notice the small one and take it with confidence. They've failed to explain their reasoning to us, but it seems that a two-nymph tandem takes more trout, especially in cloudy water, than a single-fly setup, even when no trout are caught on the larger of the two flies.

not indeed trout water in disguise. While on a trip once with some rigid anti-whitefish fishermen, I saw a flock of fish rising in a shallow, featureless run, with the tiny blips characteristic of what my companions derisively called "rubbernoses," which is actually an apt description. My friends were reluctant to let me detain them on the trip by getting out of the boat to fish over anything but trout.

We had not had many chances to fish over rising trout on the trip, and I had a craving for that kind of fishing; I didn't care what kind of fish they were. I hopped over the side of the boat while it was still moving, bounced along holding the gunwale for a few feet while my feet explored for the bottom, then let the boat go and said breathlessly, "I'll catch up with you around the next corner." I'd thought the water was shallow. They'd rather have drowned me than pull over to let me out to fish for whitefish.

I looked closely for a few seconds at the water where the fish were rising. It was about three feet deep, fairly smooth on top. A few size 16 blue-winged olive duns rode its surface. I'd been banging the banks with big nymphs. I hastily added a tapering section and two feet of 5X tippet to my leader, then tied on a small olive-bodied hen-hackled wet fly. I dressed this with floatant, cast it out, and fished it as a dry. A fish took without hesitation, with the same blip it had been making for the small naturals. It reacted violently to the hook and fought eagerly and well, taking to the air often. It was a rainbow trout just over a foot long.

I caught only three or four more, between twelve and fourteen inches long, before I had to reel up and huff off in a panic to catch up with the boat. I didn't expect my partners to wait long for a whitefish fisherman. I wasn't even sure they'd stop. When I caught up with them at the next corner and told them those rising fish were trout, they didn't believe me.

"That was whitefish water," one of them said, "and those were whitefish rises. Now get in the boat and let's go fish for some trout."

They were right in their descriptions. I'm not sure now that I even believe myself.

Reading Riffles

Defined concisely, riffles are fast and shallow water, cobbled on the bottom and choppy on the surface. They are the richest parts of freestone streams, surpassed in productivity only by weed-filled reaches of meadow streams and tailwaters. Their voices are boisterous, eager, and chatty. Riffles speak of freedom and fun and good fishing.

STRUCTURE

Riffles deliver the stream from run to run, pool to run, or pool to pool. The gradient of the streambed in a riffle is more steeply tipped than in a run, though not as steep as in a cascade or rapids. The steepness of the streambed speeds the water. The fast water does not allow deposition of silt, so riffles tend to have clean bottoms composed of rock in its various sizes.

In depth, riffles can vary from a few inches to three or four feet. Their bottoms consist of rocks that are usually on the small side, from small pebbles up to stones and boulders, with a few the size of basketballs or even bigger scattered throughout. As a result of laws of hydraulics that are not well understood yet, at least by me, the size of the rocks in a given riffle tends toward uniformity. But perfect uniformity is rarely attained; a riffle is usually broken by at least a few misfit boulders and stones in a variety of sizes. These obstruct the current and become holding lies for trout, though anglers searching for those trout cannot always see such lies through the broken surface of the riffle.

Riffles are spaced at fairly regular intervals along the course of a stream, generally at distances from five to seven stream widths apart, again in accordance with laws of hydraulics. They are fixed and constant features; riffles do not move. Some of the rocks that compose them, however, migrate individually or in herds to be deposited in the next riffle downstream during high water and are replaced by rocks bouncing down from

There is no such thing as a typical riffle, but this one, on the Yellowstone River in Montana, displays some defining features of riffles everywhere. Its rough surface reflects a bottom of boulders in fairly consistent sizes. The speed of the current is caused by the steep gradient of the river in this section relative to the water on the flat visible upstream. The depth of the water right where Jim is wading tells you there would be no trout there; they would find no boulders large enough to form shelter from the swift current, and their backs would almost be exposed to predators if they were able to hold in such brisk water.

The riffle deepens in front of Jim, where he is casting. You can also see visible signs of obstructions there. The large frothed area is obviously caused by a boulder; a few smaller areas of whitewater tell about other boulders sprinkled here and there throughout the riffle. You would expect to find trout around those obstructions, especially in the slower water downstream from them, but possibly in the pillows of soft water tucked just upstream of each. Because the potential lies would not all be marked visibly on the surface, you would want to cover all the water deep enough to give trout shelter from overhead predation, in case they're scattered on hidden lies.

the riffle above. As a consequence, many riffles are reshaped after every flood, and all riffles change shape over time. It's one reason for the famous saying, "You never step into the same river twice." It's also a reason you need to reread a riffle the first time you fish it either after a storm or in a new season.

The largest stones and boulders tend to be deposited in the surface layers of a riffle's bottom structure, seated in a surrounding of smaller rocks.

The deeper layers are composed successively of smaller stones, gravel and pebbles, and finally sand and silt as you go down into the substrate. This order of deposition results in myriad small spaces between rocks in a riffle. These spaces become what entomologist Rick Hafele calls "living rooms for aquatic insects." The larger spaces are on top, and the smaller spaces are distributed farther down. These provide a great number of places for insects to live, helping make riffles rich in trout foods.

The roughness of a riffle's bottom is a direct result of the largest rocks being deposited in the upper layers of the streambed. The migratory tendency of these large rocks means they are seldom firmly seated. This is why you sometimes feel as if you're treading on a submerged field of rolling stones when you wade a riffle. You are.

Part of the definition of a riffle is the choppiness on its surface, which reflects the roughness of its cobbled bottom. It helps you read all types of water when you understand that what you see on the surface is a reflected picture of what you can't see on the bottom. The shallower the riffle, the more clearly that picture is drawn for you.

NEEDS OF TROUT

In most streams, riffles hold the highest abundance of trout and are the first places to fish if you are after fast action. But some riffles hold trout and others don't. Those that do hold trout meet the needs of the fish consistently, allowing them to establish and stay on territories. Riffles that don't hold trout fail to meet one or another of the basic needs and are not constantly good places for trout to be. Nevertheless, such riffles might hold trout at specific times. For example, a brisk riffle that lacks obstructions to the current behind which fish can rest might become a feeding lie and hold trout temporarily during a heavy hatch of insects.

Shelter from Currents

Shelter from currents is the key need of trout in a riffle. The other needs are met with abundance here, but the ultimate quality of a trout's territory is based not on the amount of food available, but on the provision of shelter from the current and protection from predators.

Because riffle gradients are steep and their currents fast, holding lies for trout are almost always formed by some sort of obstruction to the constant flow. Though rarely as obvious as this boulder in the broad and difficult Deschutes River riffle fished by Robert Sheley, you'll increase your take of trout as you develop your ability to read both evident and less visible lies.

In this photo, Robert is reaching to drift a pair of nymphs, topped by an indicator, down the obvious line of disrupted flow below the protruding boulder. It was the only possible position from which to reach it, because the water inside it is too deep and pushy to wade. As I watched Robert wade into that precarious position from which to fish the lie, he took several trout on his way out from less visible lies on the near side of the boulder. One of them came from the slight slick almost under his rod, formed by the smaller boulder that breaks the current without protruding through the surface, right at his feet.

Most riffles have plenty of scattered stones, small boulders, and other features that break the current along the bottom, with numerous small niches behind countless small stones offering shelter to small trout. But it takes some modest-size stones and boulders in the current to hold trout of the size we don't mind catching. You'll find the kind of trout you would rather catch, bigger fish from a foot to sixteen inches long, only around larger breaks in a riffle's current. Every productive riffle has a few of these, and every one of these lies should hold a trout, its size depending on the size of the territory the obstruction forms.

The largest trout, measured in pounds rather than inches, are not common in riffles, simply because lies of the size they require are not often found in shallow, fast water. In their search for successively better territories, the largest trout leave riffles behind, though they sometimes come out to hunt in them. The exception is often found in a riffle corner, where a large trout can find its ease under the soft current inside the seam and still be in position to accept all the provender the faster current on the outside of the seam delivers.

Protection from Predators

The choppy surface of a riffle is an opaque window, like a closed curtain, to vision from above. Protection from predators is excellent. Most birds and beasts hunt their fish in slow or still water. Trout in riffles seem to enjoy a great sense of security. They are not as nervous as in slower types of water.

The broken surface of a riffle works against most predators, but it works for you and me. We can wade closer to trout without alarming them. We can fish from in close, with short casts and lots of control over the line and fly. This is important; the closer you can work to a fish in rough water, the more chance you have to present your fly correctly and to detect a take when it happens.

Trout Foods

Riffles meet the trout's need for food graciously. Two things make it so. First, all those little "living rooms" among bottom rocks prompt the growth of a great abundance of aquatic insect nymphs and larvae. Second, photosynthetic plants thrive on riffle rocks, enriched by sunlight striking down through the shallow water, giving these insects lots of rich aquatic pastures to browse. Swift riffles are constant conveyors of nymphs and larvae that get dislodged by the current. A trout waiting on its station, with its window on a slice of this conveyor, sees these insects passing and makes its living by dashing out to intercept as many of them as it can.

Aquatic insects are the primary source of food for trout in riffles, although terrestrial insects are sometimes surprised to find themselves awash in the water and taken into the drift that constantly moves past the trout. Most of the time, riffle trout feed on a mixture of various stages of a wide variety of aquatic insect species being delivered down the currents. Insects that thrive in riffles are primarily mayflies, caddisflies, and stoneflies. The nymphs, larvae, pupae, and drowned adults of these are the bits of food that riffle trout eat most often.

When you're fishing a riffle, especially a boisterous one like this that borders on pocket water, you can move quite close to trout lies before fishing them. Trout are not nearly as wary in such water as they are in calmer water. They are unable to see through the rough surface and do not fear overhead predation. The turbulence seems to make them less nervous about underwater predators as well. For whatever reason, trout do not flee at the approach of a big set of wadered legs, even though they can obviously see them coming. You can't step on trout, and you should approach slowly and quietly, but if you do, you'll be able to get right up onto trout in rough water—a rod length or two away.

Always take advantage in broken water of this shortening of what I call a trout's "worry distance," the point at which a trout's senses warn it that you've moved close enough that you're a threat to it. Terri Yamagishi is doing that here on Colorado's Big Thompson. It's much more difficult to read the most likely lies from a long cast away, and it's almost impossible to control your drift, whether with a dry fly or a nymph, if you cast much more than thirty feet. The closer you're able to approach, the more control you'll achieve. Never underestimate the value of control in all aspects of fly fishing.

Mayflies, as outlined in the chapter on trout foods, have a three-stage life cycle from nymph to dun to spinner. Mayfly nymphs that live in riffles are mostly among the flat clinger and robust crawler varieties, which are able to cling tenaciously to stones in fast currents. They live deep in the smaller niches of the substrate in their youth, where they find shelter from currents, coming out into the currents on the tops of rocks only to feed. As these insects mature and prepare for emergence, usually in spring and early summer, they move higher in the substrate and begin to feed and prowl on the surface of the riffle bottom. They get restless, they get washed into the currents, and they get taken by trout much more often.

Many mayfly species migrate out of fast water before emergence, crossing the bottom in the nymphal stage in order to emerge as duns in calmer water near shore. During these migrations, mature nymphs march in their thousands right through the territories of hungry trout. This behavior is the basis for much of the best nymph fishing we enjoy in riffles.

A few mayfly types emerge out in midriffle. In some species, the nymph swims to the surface and emergence takes place, with difficulty, right in the chop. In other species, the nymph fastens itself to bottom rocks with its tarsal claws, the chitinous nymphal skin splits open along its back, and the dun emerges there. It is then tossed at the whim of the currents on its wild swim to the surface. When a lot of these mayflies come off in a riffle at the same time, in a mass emergence that depends on sheer numbers for survival of the species, it's an obvious cause for excitement among trout. It appears to be a time for dry-fly fishing, because what you see, with your view only of the top, is a flock of helpless duns riding the bumpy surface. But a wet fly or nymph fished a foot or so deep mimics the rising nymph or dun and will often take more trout than a dry.

Once the duns reach the surface, they must ride it out until their wings dry and they can fly. Because the surface is rough, many drown and become part of the swirling drift. Again, you see duns on top and it seems like dry-fly time. You should obviously try a dry, but wet flies are often better.

Duns that survive the rough surface and manage to get airborne fly directly to streamside vegetation, where they molt into spinners before returning to mate in the air and lay their eggs on the water. Most egg laying takes place over runs and flats, not on riffles, though some spinners lay their eggs in riffles. Many spinners that lay their eggs on calmer water fall to the water, drown, and are delivered downstream to riffles as a major component of the drift.

Riffle-dwelling caddisfly larvae are primarily free-living green rock worms or net spinners in approximately the same shape and color. They don't build cases. They are wormlike, cannot swim, and are frequently dislodged by the current for the convenience of foraging trout. Their presence in good numbers in a kick-net collection sample is an indication that nymphing the bottom will be an excellent way to fool fish.

When their full growth is achieved, caddis larvae seal themselves into submerged cocoons, and the pupal transformation into the pharate adult

During a workshop that I gave recently for the Fidalgo Fly Fishers in Mount Vernon, Washington, the discussion ran toward reading water. Ecologist Bruce Freet made the sensible observation that in all of life, the most productive areas are along edges. You find game, big and small, at the edges of forests. You find game birds at the edges of fields and cover. You find ocean fish at the edges of currents. Not all animals are located at the edges of things, but most edges of things concentrate the numbers of animals of all sizes along them. If prey is gathered along edges, it's certain that predators will gather there as well. Trout are predators.

Streams have all sorts of edges: bank water, seams, the edges of riffles and runs, edges formed by overhanging vegetation, more edges formed by rooted aquatic vegetation. All of these concentrate trout foods and therefore attract trout.

In this photo, it's easy to read the edge of the riffle on Chile's Cisnes River, being fished by Marcelo Dufflocq under the eye of John Eustice. Marcelo is called "the Cormorant" for his ability, since his youth at the lodge owned by his father and mother, to find and catch trout. It's no accident that he is fishing right up the edge of the line of current drawn down that riffle. It's an easy one to read, and trout will be concentrated directly under that line.

form takes place safely inside. When ready, they scissor out and make their way to the surface. In some species, emergence is en masse, as with most mayflies, which can cause selective feeding. In many caddis species, emergence is scattered. They come off in a trickle, which can keep riffle trout on the lookout for caddis pupae and adults over long periods of the day and the season.

Caddis pupae are among the most exciting, but least visible, aquatic insects in riffles. Though some can swim and are also buoyed up by gases, for the most part they are at the whim of the current until they reach the surface. But a riffle is shallow; it's a brief transition from bottom to top, and trout know it. They rush caddis pupae eagerly, often making the successful interception just beneath the surface, causing what the late Gary LaFontaine, in his book *Caddisflies*, called "pyramid rises." These rises send little spurts of water into the air. They look like takes to floating caddis adults but usually indicate the death of a caddis pupa instead.

Sometimes a trout chasing a pupa to the surface lofts gracefully into the air above the riffle, arcs over, and arrows back in, propelled by its own momentum. This odd and beautiful behavior is usually seen just at dusk, so the arcing trout are black silhouettes against the brighter water. It almost always indicates trout feeding on the pupal stage of the caddis, not the adult.

Caddis emergences in riffles can be confusing because what we see, from our viewpoint above the water, are not pupal caddis, but escaping adults that have survived the trip to the top and shot off into the air. Trout feed on pupae, but their swirls and splashes are at the surface, and they appear to be feeding on adults. In my own past, I used to try every dry fly I owned in these instances, while the trout fed greedily just under the surface and ignored them all. I've learned that switching to a subsurface nymph or wet fly, approximately the size and color of the visible adult, will usually do the trick and take at least a few of the trout. I often drop the sunk fly right off the hook bend of the dry I've been trying, on two feet of tippet, in a dry-and-dropper rig. This gives the trout a choice between adult and pupa patterns.

Even when caddis are not emerging, a lot of caddis adults are still out and about, busy bouncing above riffles, laying eggs, and getting eaten by trout. Riffle trout get used to seeing the adult caddis shape and seem quite eager to spear upward out of their shallow holds to take it, or something that imitates it. Two of the most successful dry flies I have ever used for riffle fishing, the Elk Hair Caddis and Deer Hair Caddis, are based on the shape of the caddisfly adult.

Stonefly nymphs, though numerous in riffles, are hidden insects that live in niches among the stones. When ready to emerge, they crawl in their migrations toward shore. If you do not sample the bottom by lifting rocks to see what might be clinging to them or taking kick-net samples, you might never know they're there.

Trout seldom feed selectively on smaller stonefly nymphs, but they do eat a lot of them, usually as individual items picked up along the buffet line that is the riffle's current. These add to the various mayfly nymphs, caddisfly larvae, and other unhappy creatures that make up the drift on which riffle trout feed most of the time. They increase the reasons that bumping a small to medium-size generic nymph, such as a weighted Gold-Ribbed Hare's Ear, along the bottom is an effective tactic in a riffle.

Migrations of large stonefly nymphs, such as salmon flies and golden stones, are exceptions to the nonselective rule. These are so large, size 4 to 8, and move in such large numbers that trout become selective to them, even in riffles, and you need to imitate them.

Because stoneflies crawl out to emerge, usually at night but sometimes on cloudy late afternoons, their actual hatches are seldom events that interest any but sleepless anglers. But many stonefly adults deposit their eggs over riffles in afternoon and evening, and these adults are an important source of food. They don't often cause trout in riffles to feed selectively, but they do cause trout to turn their attention toward the surface, so you can entice them up to a dry fly. If the stoneflies depositing eggs are salmon flies or golden stones, trout will reverse their nonselectivity rules for riffles and focus their feeding on the large insects, often taking them with detonations that you can hear for fifty yards over the rushing sound of the riffle from which the trout arose.

Mayflies, caddisflies, and stoneflies make up the bulk of the diet for riffle trout. Other creatures find themselves inserted into the menu from time to time, including cranefly and midge larvae; various terrestrials such as beetles, hoppers, and ants; and a modest amount of annelids, which are aquatic worms. These are taken most often as drifting snacks rather than main meals, but that is the nature of most trout feeding in riffles: a bite of this and another of that, with selectivity only occasionally getting into the mix.

Temperature and Oxygen Levels

Riffles are rich in oxygen, second only to frothed rapids and cascades, where the gradient of the streambed is tipped even steeper. Temperature and oxygen levels are interrelated, with cooler water containing more oxygen, and warmer water less. When rivers suffer summer low flows, and fish begin to get distressed by warm water and low oxygen levels, trout move to water with the most oxygen. In the absence of springheads or feeder streams, trout will often hang in riffles, if they can find suitable lies in

When conditions become warm along any stream, oxygen can become depleted, because water entrains less oxygen as it warms up. Because water is aerated in riffles, trout will nose up toward them or hold in calm pockets within them if slack water stresses them. It's important that trout already under stress from stream conditions be played out quickly and released. In extreme conditions, you should not fish for them at all.

which to escape the fast current, or in the nearest holding water below a riffle, to take advantage of oxygen injected into the stream through the broken surface.

If conditions are extreme and you find no trout holding in riffles, your next logical choices are to fish pocket water or explore until you discover cool springs or shaded headwaters. Sometimes trout move into tiny creeks and you'll find them there in sizes and numbers out of proportion to what the creek would be expected to provide. Fish for them there only if their condition is good enough to survive a fight with you.

Whenever you hook a trout that is already stressed by high water temperatures and low oxygen levels, you risk its life. Land it quickly and release it gently. If conditions are marginal for trout survival, you should not fish at all.

Holding lies in riffles, as opposed to feeding lies or empty water, have at least a few breaks in the current where trout can set up stations and defend territories. One of the major dimensions of any territory is its depth, and riffles are always limited in that dimension. The primary food-producing area of any territory is its bottom, and secondary is its surface, but midlevel drift can be significant as well. A territory set up in shallow water is not going to be as large in volume as a territory of the same size in deeper water.

Because the size of a trout is relative to the quality and size of its territory, even a riffle that has good holding lies throughout its length usually holds trout somewhat smaller than the largest trout that can be caught from deeper water types in the same stream. But their rich aquatic life combined with the high number of compact niches make riffles good places to start a day on any stream. If you are not a bounty hunter after trophies only, riffles might be good places to spend your day.

You can catch lots of trout in riffles. All will be feisty, and many will be nice ones.

Empty Water

Pinpointing potential holding lies in a riffle and finding its trout are relatively easy. The first step is to eliminate obvious empty water. On paper, this means water only a few inches deep, too thin to protect trout. It also means dead water off to the sides of the riffle, where no current delivers food. You won't find trout in still shallows that taper up onto gravel bars, where kingfishers, ospreys, or herons would easily spot and attack any fish. You can eliminate an entire riffle if it is too shallow and unfeatured, lacks any obstructions to break the current, or flows so swiftly that trout could never hold anywhere in there.

On the stream, as opposed to paper, empty water should be dismissed with a little more caution. Fly-fishing literature is full of scoldings for those of us who assume a certain kind of water would never hold trout, always by those who have caught rare whoppers in it. Scan the skinny water, especially if it's bumpy and hard to peer into, to see if it might not hide feeding trout or have hidden features that would shelter trout and form permanent lies. Look for signs of a hatch; a few insect types migrate right to shore before emerging, and terrestrials such as beetles and ants fall to the water there. Trout sometimes follow and risk feeding on them in water where they would not otherwise be found because of danger from overhead.

Recognition of empty water in riffles will cut back on time wasted fishing it and increase the amount of time you fish water with potential to provide trout. Though the water in which Jim Schollmeyer wades is rich in terms of insect production, all that richness will be delivered over the shelf and into the deeper water downstream. Trout in this case would clearly avoid the water that is too fast, chattery, and shallow to hold in and concentrate in the nearest slower and deeper water downstream from it. They would be tucked up against that shelf, probably all along it, but in greatest numbers where the slight slicks on the surface denote calm waters on the bottom. Those are prime lies.

You can usually declare up to half of any riffle to be empty water. If you've defined it correctly, and refrain from wasting time fishing it, you automatically increase your catch, because you spend your time casting where you're highly likely to find trout.

Once you've eliminated empty water, it's time to examine the remaining potential holding water in a riffle to decide where trout are most likely to be found in it. There are several obvious places to search with both your eyes and your flies. You're looking for high-percentage water, and it's the way water meets the needs of trout that propels it into that category.

Since most of a riffle offers protection from predators, as well as plenty of food, shelter from currents becomes the obvious thing to look for when you are trying to find holding lies in fast and shallow water. Any part of a riffle that satisfies all three needs of trout will be a prime lie and is likely to be the spot where you'll find the largest trout in the riffle.

Corners

A corner is where a riffle breaks around a point of land and forms a ledge cutting at an angle slightly to almost directly across and a seam running lengthwise, and it's the first obvious place to look for holding lies. If a stream has riffles, it will have riffle corners, though not every corner has the right shape to serve as a holding lie for trout. All freestone trout streams have riffle corners. Even meadow streams form riffle corners in many of their bends—not surprisingly, where they turn a corner.

Riffle corners are the most likely places, in any stream that has them, to find a few trout. Though corners often form at the head of a riffle, just as many form where a point juts into the middle of the riffle or where the river, whether freestone or meadow, makes a bend. The point that forms a corner is almost always a gravel bar; that's the bottom of the riffle when the water is higher.

If you spend any time watching fishing guides work, which is not a bad way to improve your ability to read rivers, you'll see them ease their drift boats down through pools and runs and rapids, encouraging their clients to cast while they move. But they will almost always pull over, beach the boat, and fish their clients carefully at the corners of riffles.

It's not always the top end of the riffle that should draw your attention, but the slight eddies formed at any corner. These are wedges of water squeezed between the rough riffle on the outside and the still water against the shore on the inside. They are always triangles, little larger than a yachtsman's flag on tiny riffles, rarely wider than a short cast across even on large riffles in big rivers.

One backhand way to describe this prime water type: It's where you'll always see somebody wade right through the good water, scattering trout in all directions, to begin fishing the deeper part of the riffle out beyond it, where the water looks a lot more fishy but holds a lot fewer fish. Never wade into any riffle corner until you've popped at least a few casts to water so shallow that it looks as though it's empty water that shouldn't hold any trout.

The upper end of a riffle corner is usually just a foot or two deep. It gets deeper as you follow the seam downstream. The most productive water in a corner is at least slightly rough on top, but not as choppy as the riffle outside it and not as smooth as the backed-up water to the inside. It's

Riffle corners form prime lies in streams of all sizes. This riffle, on Kettle Creek in Pennsylvania, turns in the open for part of its length, and that would all be productive water. But the upper part of it dives against an undercut bank and forms the perfect lie for a nice trout, which Jay Nichols is trying to prod out with a dry-fly cast short and fed into a downstream drift.

a seam, a transitional line of water down along the edge of the riffle, that varies from a foot wide on small water to about fifteen feet wide on big water. The productive seam might extend for just a few feet downstream from the very corner or follow the riffle for fifty feet or more.

The reason a good riffle corner is such a prime lie is twofold. First, the riffle upstream from the corner is extremely productive, conveying a constant supply of aquatic insects, but provides few features where trout can find shelter from the current. Second, the inside of the seam formed by the corner is the first soft spot where trout can hold, avoiding the force of the current, but be in position to rush out and intercept what that same current constantly delivers. That's why trout line up along the length of the seam. It's also why the largest trout, seeking the best territory, often tucks itself right at the very head of the corner, where it can get at the groceries first.

One fall, I was fishing with Jim Schollmeyer, fishing photographer and coauthor, with Ted Leeson, of *The Fly Tier's Benchside Reference*. We parked our boat alongside a long riffle on the Bighorn River in Montana. I

This big corner, on the Yellowstone River below Big Timber, Montana, has a very long seam where the riffle tails out into a run. It's an excellent example of edge water, but big trout might also hold all across the water downstream from the break, because they would be protected from overhead predation and find plenty of drift on which to feed. The small trout Jim has lofted on his dry fly, however, struck along the seam just downstream from the corner. Often the closer to the corner you get, the larger the trout you catch, because the lie becomes more prime as you move closer to the source of food.

started fishing the seam between fast water and slow about thirty feet below the head of the riffle, where it formed a perfect corner. As I slowly worked my way up toward the corner, without drawing anything to my Elk Hair Caddis dry, I felt building pressure to provide Jim a trout for a photograph. He idled on the bank behind me, resting his camera on a tripod while waiting for me to hook a fish. I eventually cast almost into the corner, where it shallowed out and looked empty to me, without a take. I began to get ready to quit.

From his higher vantage point on the bank, Jim spotted a trout rising. "Farther up!" he said.

I thought I had already fished the corner out to its upper end, but I moved up and took a couple more casts to where it seemed little water was left. Nothing rose to the fly, and I looked up at Jim to get more directions. "Farther up!" he ordered again. "Farther up!"

This riffle corner, on Oregon's Willamette River, would benefit from the enormous productivity of the shallow water upstream from the shelf, but that water would be empty of trout. As with all riffle corners, fish would tuck themselves up into the corner, the largest ones first, in order to be first to the food delivered downstream from above.

I made a cast to where the corner was about the size of an arrowhead point. A trout took with a subtle swirl, felt the hook, and pounced out into the current. It used the force of the river to drive it, and I was obliged to splash after it. Jim snapped pictures when I finally led it into quiet water two hundred feet down the riffle. It was a rainbow, about three pounds, but it fought far above its weight class.

We talked idly while we walked the gravel bar back toward the head of the riffle, Jim carrying his camera and tripod slung over his shoulder, me carrying my fly rod slung over mine. I was going to stop at the boat. I'd provided that trout; it was time to float down and join friends for lunch.

Jim said, "Take another cast up there," and aimed me back at the corner.

"What for?" I asked.

"Just in case" was his reply.

I cast. Another trout took with the same subtle swirl and charged off on the same river-propelled flight. We wound up landing it in the same spot as the first fish. Jim took the same pictures all over again, and I released the trout. This one weighed closer to four pounds.

Trout take up stations along the length of the seam, right up into the corner itself. They also sprinkle themselves throughout the softer water just down-stream from the shelf that cuts across stream, interrupts the riffle, and provides relief from the descending current while protecting trout from overhead preda-tion and furnishing a constant supply of insects in the drift.

Those were uncommonly large trout to be holding in a riffle, even at the corner, but the Bighorn is not your common river. Never neglect the corner of a riffle on any river.

The Lower End or Tailout

A lot of life is produced in the length of a rich riffle. A lot of that life is dis-lodged daily. The lower down the riffle you look, the more chance you'll find bits of life adrift on the current. If the riffle has obstructions throughout its length, trout will hold wherever they can, sorted by size according to the

If a riffle is too pushy and has too few obstructions to break its fast flow, trout will not find any places to hold in it. They'll hold in the nearest modest currents that offer them security and nose up into the tailout of the riffle whenever it provides abundant food. In this situation on the Clearwater River in Idaho, trout would not be able to hold in the broad and brisk riffle, so they would position themselves in the upper arc of the slower and deeper water just downstream from it. They would find plenty of depth to protect them from predators there, though the camera angle makes it look shallow. They would also find plenty of places along the bottom where the current is broken by all those boulders.

amount of shelter provided from the current. But a riffle that has no places where trout can hold will be used only as a feeding riffle or not at all.

If a featureless feeding riffle of this sort is less than boisterous, trout move up into it to feed whenever insects are active and the drift is plentiful. If the current is fast, however, trout tire in it quickly and are unable to hold in it even to feed. Rather than moving into it, they hold where the riffle breaks at its lower end and wait for the current to deliver drift farther down, where they can hold stations in some comfort. This is transitional water, sometimes a shallow sort of tailout where the riffle shelves up a bit, sometimes an easing out where the riffle merges into whatever kind of water comes next, a run or pool.

If water near the tail of a riffle offers trout shelter from currents and protection from predators, and if a hatch or some other insect activity is going on in the length of the riffle, then trout will often move up into the lower end, remaining there to feed only so long as enough food is available to keep them ahead in the energy equation. When the food supply dwindles, trout drop back down to their stations in nearby water.

Obstructions

It should by now be obvious that the key to holding water within any riffle is an obstruction that slows the current. But because a riffle tends to accumulate similar-size rocks and conceal them beneath a uniformly choppy surface, lies formed by obstructions are not always visible to the angler's eye. These are the kinds of lies Ray Bergman referred to in his famous 1938 book, *Trout*, when he said about riffles, "I wager that nine-tenths of the anglers skip them, considering them unworthy of notice. Some of them are—no doubt about it. But others have from one to ten *pocket holes* that contain fish." Those holes are formed by invisible obstructions.

The most common obstruction that forms a holding lie in a riffle is a single boulder. It breaks the flow, leaving a calm area that is the width of the boulder at the upper end and slowly narrows until the split currents rejoin downstream from the boulder. The second most common kind of obstruction is a small gathering of stones slightly larger than the rocks on the riffle bed around them. These deflect the current in several directions, causing turbulence, amid which there is usually a place where the water is slower. This becomes a trout's station, and the water around it is the trout's territory.

These kinds of stations, ranging from small ones that hold tiny trout up to larger ones that sometimes, though far from always, show on the surface and usually hold larger trout, are scattered throughout the length of most riffles. They're the main reason you should cover all the water when you fish a riffle that looks productive, rather than just placing your casts to the most promising places. If you do that, you'll not make contact with trout holding on all those invisible lies.

Many prime lies in a riffle are marked on its surface and easy to spot. The most obvious among these are any boulders that either break the surface or are sufficiently close to cause a boil on top. Any boulder large enough to disrupt the surface is large enough to shelter at least one trout from the current, sometimes two or more.

Water pillows in front of a boulder, and this creates a soft spot in the current just as effectively as the eddied water behind it. Because the pillow of slow water gives a trout an upcurrent view that is unobstructed by the boulder, it's often a more efficient place for a trout to lie on station, surveying its territory for bits of drift, than a lie downstream from the same boulder.

The Madison River is often described as a "fifty-mile riffle." It's apt; the most heavily fished part of the river between Ennis and Quake Lake has a few runs and what might be called pools, but most of it has a gradient too steep and constant for anything but riffles to form. Because it rarely levels off like most rivers do, the Madison offers few riffle corners. A high percentage of lies on the river are defined by boulders, many of which are easy to spot because they either protrude from the surface or send up a white wave where the water plunges over them, much like a frothed breaker on an ocean beach. The river is studded with many more boulders that are invisible along the bottom or marked only by slight boils on the surface. All are excellent lies for trout that are very often large.

When you float the Madison with a guide, you'll be urged to cast a few feet upstream from visible boulders so that your fly sinks and swings into the soft pillow of water that forms where it divides to push around them. In my own minor experience floating and fishing the Madison, the largest trout have always been taken from the upstream, rather than downstream, lies built around boulders.

Though fishing literature is full of instructions to fish thoroughly in the broken water downstream from midstream boulders, I have taken as many trout from the water immediately upstream from them. Don't neglect the water below, but always be sure that if a fish is holding upstream from a boulder or any other obstruction in the current, it gets at least a fair look at your fly.

Ledges, Trenches, and Shelves

Ledges and trenches break the current and form lies for trout. They are generally structures of base-rock bottoms and are not common in riffles, which have cobbled bottoms. But some streams have riffles that work their way to bedrock in a few places, and the water erodes shelves into shale and leaves undercuts that trout can creep under. A base-rock ledge or trench in the middle of a riffle will show as a smooth slick in the choppy surface.

In some riffles with cobbled rock bottoms, a lodged boulder or a small grouping of them, with their tops even with the streambed, will cause a trench to erode into the rocky bottom just downstream. Such trenches form lies for at least one trout, sometimes a pod of them, because the current rushes overhead.

Any trench formed in a riffle is a potential prime lie. It offers a shield against the current. Its depth provides overhead protection from predators. And the riffle, with its richness of insect life, delivers food that magically drops into the trench. If any riffle holds a trout that you would weigh rather than measure, this is the kind of lie where you might find it.

The Nez Perce Flats reach of the Firehole River in Yellowstone Park has lots of such lies. This is a riffled section alternating with flats, most of it one to three feet deep. Volcanic bedrock lies just beneath the rocks that form the riffles, and this bedrock is pocked with depressions a foot or two deeper than the rest of the water. Most of these ledges, trenches, and shelves are recorded on the surface as slight slicks.

Fishing these heavily pestered Firehole riffles is easy once you catch on to wading downstream, rather than upstream as most folks do, dropping a dry ahead of you onto the water above the slicks, feeding slack for a downstream and not upstream drift. It also works well to cast a wet or small nymph above and beyond these lies, letting the fly sweep slowly down and across them on the current. Sometimes you'll see a swirl as a fish takes in a patch of smooth water; other times you'll feel your line tighten slowly, then feel a sullen pull before you raise the rod to set the hook in a trout that intercepted your wet fly or nymph as it crossed above its holding lie.

In many riffles, the gravel suddenly shelves off, cutting laterally across the stream. The drop is never as abrupt as in bedrock, but the sides slope steeply into deeper water. Probably the best way to describe one of these is to remind you of the time you waded a shallow riffle without a care, then took one more step and suddenly felt the gravel go out from under you. Backpedaling furiously, you managed to save yourself. What you kept yourself from falling into was a fine trout lie, or even a line of fine lies, at the foot of a sloping gravel shelf.

A slick down the length of a riffle tells about one of two things on its bottom: a boulder at its head or a trench cut the length of the slick. If it's a boulder, you'll almost always see signs of a boil at the upper end and turbulence downstream from it, where the separated currents slowly rejoin.

A slick that forms with no boil or turbulence usually denotes a trench cut into the bottom. These are especially productive, because the current is broken along the streambed there, and trout don't have to fight the flow. The trench is always a few inches to a foot or more deeper than the water around it, so it offers more protection from predators than the surrounding riffle. Food is no more and no less abundant in the trench, but it does tumble in and becomes a lot easier to capture. So trout in a trench cut into a riffle find all their needs met in that one spot, which makes it a prime lie.

Don't ever neglect the water on both sides of a trench. But do learn to recognize them by the slicks they form, slight to very visible, and focus your fishing so that your fly either drifts just over them or tumbles into them like natural food forms.

These gravel shelves usually run at angles across the current, though some run parallel to it. They are among the finest trout lies in riffles and hold some of the largest trout found in them. The shelf breaks the current, and the depth of the water gives some protection from predators. The swirl of the current downstream from the shelf causes trout food to gather, which makes it easy for trout to get fed. Trout take stations all along such a shelf, their noses sometimes poked tight up against it.

Riffles, such as this one on the Firehole River in Yellowstone Park, hold trout in many scattered lies that are not always visible on the surface. It's often best to fish them with a wet fly or nymph on the swing, swimming the sunk fly through all of the water, therefore showing it to trout in all potential lies.

Shelves usually show up on the surface as lines where the riffle's wavelets are suddenly and obviously different than they are throughout the rest of the riffle. The difference is usually easy to see, and you should have no trouble picking them out. Sometimes small shelves cause no more than a minor depression in the bottom. These are more difficult to spot, showing up only as areas where the chop is a bit higher or lower than the chop around it or as lines of slightly darker water on the surface of the riffle.

Even small depressions in a riffle, whether formed by a ledge, trench, or shelf, hold trout. Always fish them carefully.

Current Seams

Current seams are the sewing together of two currents of different directions or speeds. Wherever this happens in a riffle, feed delivered by the separate currents gathers along one line, and a trout finds it a fine feeding lie—or a bunch of trout find it a fine line of lies. If the currents conflict enough to cause turbulence, and therefore some pockets of consistently subdued flow in riffle water that otherwise rushes right along, trout will establish holding lies on stations beneath a seam along the bottom.

Recognizing seams becomes such a sixth sense that an experienced fly fisherman will angle casts along a slight seam in a riffle without even noticing he's done it. Yet when he stands in the riffle and tries to point out to you what demarks the line he fishes, he might not be able to do it. The difference is usually in the chop on the surface. Wherever you see a line of water with wavelets slightly sharper, taller or shorter, or just more agitated than those alongside it, consider it a seam and a likely place to hold trout. Fish the entire length of the seam, placing your casts so that the fly, whether a dry fly or nymph fished upstream or a wet fly or streamer fished down and across stream, covers the centerline of the seam and the water on both sides of it.

The seam along the side of a riffle, between fast midcurrent flows and slow edge-current flows, is the easiest one to see. It is often, though not always, a likely line of holding lies. It will have the most potential in riffles where the water on the inside is a foot or two feet deep and somewhat slow, and the water of the passing riffle is two to four feet deep and much faster. The seam between fast and slow water is clearly defined, with the smooth water to the side abutted against water out in the riffle that is choppy, and also often demarked by a change in color because of the sudden acquiring of some depth. Such edge seams should always be fished carefully, if you read them as being likely lies, not so shallow under the seam or so fast right to the edge that they are more likely to be empty water. Trout often hang along the entire length of an edge seam in a productive riffle, on the bottom under the soft side, where they can dash out to intercept feed drifting on the fast side. Again, any obvious obstruction under such a seam might denote a prime lie.

Current seams always form where separated currents rejoin. They might be split by an island, a shallow gravel bar, a boulder in a riffle, a log lodged in it, or a shelf cutting across it. The seam downstream from a riffle corner forms a classic line of holding lies. No matter what the cause of a seam or how obvious or obscured it might be, always explore its length with your eyes, looking for rises and suspended feeding trout, and your flies.

FISHING STRATEGIES

Selectivity is not usually an aspect of trout behavior in riffles. They feed at random from the drift most of the time. Their posture on their stations is one of readiness to move out and intercept bits of food. Their focus will be on whatever water level has been providing the most grub. Most of the time that's the bottom, but a riffle is shallow, and trout are often attuned to drift in the mid-depths or on top.

Though most anglers turn instantly to dry flies when fishing riffles, it's best to think in terms of the three levels at which trout feed: the bottom, mid-depths, and top. Even though the water is shallow, trout with their attention focused on one of the levels will sometimes ignore flies presented at any of the others.

Tackle

Tackle for fishing the three levels of a riffle is all the same, unless you happen to command a caddy and can shout him up with another rod at the slightest change in the situation.

It's usual to choose your armament based on the size water you're fishing for a day, not the water type you're fishing for a few minutes as you move from place to place along the creek, stream, or river. On a creek or small stream, for me, that will almost always be a 7- to 8-food rod, balanced to a 3- or 4-weight floating line. The rod is short to avoid overhanging tree branches. The line is light, but not so light it won't command hackled size 12 and 14 dry flies, which I cast most often on small water. Rods balanced to 1- and 2-weight lines are too soft, cast loops that are too open, for my taste on tiny water. I like brisk actions that throw tight loops and place my flies with great accuracy to hit the small targets that are the only kind small streams seem always to offer.

On a medium-size stream or small river, I'll usually be armed with an 8½- to 9-foot rod balanced to a 4- or 5-weight floating line. Obstructions overhead are rarely a factor on such water, but control of the cast after the line is on the water becomes important, and the extra rod length offers that. The rod will still have a medium-fast to fast action for fairly tight loops when I'm casting dry flies. But I don't like a rod so fast that I get nothing but tangles when I switch to an indicator, a split shot or two, and a nymph or two. Tight loops with such complicated terminal gear will keep you busy unsnarling yourself. Again, this is the rod I'll be carrying as I move along the stream, and I'm not likely to head back to the rig to get another if I encounter a water type where it's not perfect.

I most often use the same rod on a large river. There is little difference between fishing a riffle on an average-size trout stream and another on a big river, except the latter might require longer casts. I always want to be prepared to fish either hatches on top or nymphs along the bottom on any size stream or river, and I find that the same rod, chosen correctly, prepares me for either.

If I were planning to fish a large river with nothing but large nymphs or streamers, then I would carry an 8½- to 9½-foot rod balanced to a 6-weight

Wherever a riffle feeds into a current with chop on top and a seam along the side, you should probe that seam with whatever type of fly or flies you're fishing, surface or sunk.

line, with some pretty stiff propellant in the stick. But I would almost always be fishing these kinds of flies from a drift boat, in which case I simply carry both the medium and heavy rods in the boat, strung and ready to fish. If I'm wading, I'll choose; I'm not going to carry two rods.

A leader about the length of the rod is all you need in a riffle. I buy 7½- and 9-foot tapered leaders, with tips the size I intend to use and one size stouter, usually 3X and 4X. I carry spare tippet spools in 3X to 6X. On the stream, if needed, which usually is the case, I can add tapering sections and a tippet of the diameter I want and fish any situation. Whatever leader you use, it should gracefully turn over the flies you cast with it.

Fishing the Bottom

The best approach for fishing the bottom of a riffle is with the upstream nymph technique, rigged with a strike indicator, split shot, and one or two nymphs. Begin fishing from the lower end of the riffle, cover all of the water with short casts, and work your way slowly toward the upper end. If the water is a couple feet or more deep and looks as if it might have hidden lies throughout its length, set up a pattern to cover all the water. If you read

The inside edge of a big riffle, such as the one fished here by famous fly-fishing author Skip Morris, will often have trout lined up the length of it. They hold in the slower water on the soft side of the seam and position themselves to dash into the faster current to take whatever it might deliver past them. Skip is fishing a streamer in a manner that swims it from the fast water across the seam into the slower water.

most of the riffle as empty water but notice a few obvious holding lies, high-grade the riffle by fishing just those lies.

Your nymph selection for riffles should be related to the natural insects that dwell on the bottom, though it's not necessary that they be imitations. For the most part, the insects will be mayfly nymphs, stonefly nymphs, and free-living caddisfly larvae. They run from tiny size 20s up to size 4s. On the average, though, they run in the midrange of hook sizes, from 12 to 16, and it's this average that trout see and eat most of the time. You should generally use nymphs smaller than you think would be appropriate for the water. In the absence of some compelling reason to use a large nymph, such as the overwhelming presence of big salmon fly nymphs, stick with size 14s and smaller. If you rig with a two-fly tandem, use a large or medium as the top nymph, but the other should be size 16 or smaller. You'll usually catch most of your trout on the smaller of the two nymphs.

Natural nymphs blend with bottom stones for camouflage, or they'd quickly have become extinct. The colors of the most common aquatics run to dark grays, olives, and browns. Your nymph selection should reflect those colors.

A refined fishing vest, bulging with fly boxes, will house an incredible collection of imitative nymphs. But a few general searching nymphs are usually most effective in riffles. Don't do as I did when I began my fly-fishing career: I matched all the rarities and oddities I found and wound up with a tangle of dressings that imitated creatures I had seen only once, and most trout had never seen at all.

To cover the size and color spectrum of most naturals, you need to carry only a few Gray Nymphs or Muskrats, Gold-Ribbed Hare's Ears, Dave Whitlock's Red Fox Squirrel Nymph, and peacock-bodied Zug Bugs or Prince Nymphs in sizes 12 through 16. Some should have bead heads. Broaden your selection with a few Copper Johns and Lightning Bugs in the same sizes. If you already have favorites, take them along and fish them first. Your nymphs should be slightly weighted, and you should also carry a selection of small nonlead split shot or putty weight to get them down to the bottom when their internal weight is not enough. The speed and depth of the water will dictate the amount of weight needed.

A bright strike indicator on the leader will help you spot the hesitation that marks the take of a nymph in a riffle. The indicator should be above the top nymph a distance of one and a half to two times the depth of the water. This allows the nymph a natural drift near the bottom, while keeping the indicator on the surface where you can see it. The indicator can be a bit of orange or yellow yarn tied into the leader, a large and bushy dry fly, or any of the commercial cork or foam indicators available at fly shops.

Few natural nymphs and larvae that live in riffles swim well. When dislodged from their holds on the bottom, they tumble along with the current until they're able to regain their footing. You want your nymph to perform approximately the same maneuvers, tumbling along in what is commonly known as a "dead-drift presentation" but along the bottom, just as a dry fly might float freely on the surface. In order to achieve that, you're almost forced to fish upstream.

Start by casting almost straight upstream, then make each successive cast a foot or two farther out into the riffle. Continue until you've worked all the water you can from the initial position without casting more than thirty-five to forty feet. Keep your casts shorter if the water and your self-discipline allow it. After you've covered the water you can reach with short casts from your first position, wade upstream a few feet and repeat the pattern.

If the riffle has potential to hold trout anywhere, it's best to fish prime lies only as you come to them, fishing all the water in the entire riffle around them. Then fish the prime lie thoroughly, working the nymph above and below boulders, deep into trenches, along ledges, and down on the bottom beneath any seams where two currents meet. Read the water for the most likely holding lies, and present your nymphs most diligently where trout are most likely to hold.

As you fish out each cast, hold your rod tip high enough to keep as much line off the water as possible, but not so high that you have no reach left with which to set the hook. Follow the indicator downstream with the rod, drawing in slack as the indicator and nymph drift toward you, tossing slight mends to align the line and leader with the indicator, and dropping the rod tip and feeding out slack as the nymph passes you and drifts downstream. Extend the drift as far as you can. It takes several feet of drift to get your nymph to the bottom; keep it there as long as possible. When the rig has reached the end of its drift, let the current lift the nymph up from the bottom. Strikes often occur at this point, though you'll miss a lot of them because in setting the hook, you draw it away from the trout rather than into its jaw.

The strike indicator should ride on the surface or just an inch or two below, where you have no trouble seeing it. Because the nymph should be hitting bottom from time to time, your indicator will give you false reports on many drifts. It takes some practice to get a sense for what is a take and what is not. The best rule is this: If the indicator hesitates at all, or moves in any way contrary to the current, set the hook. Sometimes it will be a fish, sometimes not. But if you don't set the hook, you'll never know.

At times the indicator will take a sudden bounce upstream or an abrupt dip down, spelling an obvious interception. Most takes are more subtle. The trout merely noses over from its station, flares its gills to inhale the nymph, mouths it, doesn't like the taste or texture, and closes its gills forcefully to reject it. Meanwhile, the indicator has hesitated. It can be almost imperceptible. The surprising thing in this kind of shot and indicator nymphing is the number of trout you'll hook in riffles once you get the hang of it. As you become more expert at it, you'll rarely be able to explain just what it was that triggered your hook set. When setting the hook becomes a sixth sense, you have upstream nymph fishing down.

Fishing the Mid-depths

You can fish the mid-depths in riffles with the same set of nymphs you use for fishing the bottom, but without any weight added to the leader. Most of the insects that trout find adrift in the mid-depths, however, are emerging mayfly nymphs or caddis pupae, with their legs and antennae

Trout on a big and broad riffle such as this one fished by Rick Hafele, on the Deschutes River, might be sprinkled anywhere across it, but are most likely to hold in association with boulders, which break the current. Though a single trout might be tucked into the pillow of slowed water upstream from each boulder, most will be in the calmer currents downstream from them.

and other peripheral parts trailing in the currents, or drowned terrestrials and winged aquatic insect adults. These have more moving parts than natural nymphs and are perfectly imitated by either traditional or simple soft-hackled wet flies.

Patterns that work best for me in riffles are standard wets, such as Light and Dark Cahills, Hare's Ear and Leadwing Coachman, and soft-hackles such as the Partridge and Green, Partridge and Yellow, and March Brown Spider. Typical sizes are the same as for nymphs used to fish the bottom in the same water: 12 through 16.

Rig by tying the wet fly to the end of your tippet. The line should be a floater. If you're already fishing a dry fly, nip it off and tie on the wet. If you want to fish a pair of wets, add a foot or two of tippet to the hook bend of the first fly, and tie the second to it. It's that simple.

To fish mid-depths with a wet fly, step in at the head of the riffle, cast slightly downstream from straight across, and allow the fly to swing around slowly with the current. If the fly starts to swing too fast, slow it by mending the line, lifting up the downstream curve of the line and looping it over upstream. Strive to achieve a drift in which the line leads the fly like a poodle on a leash, without dragging it. This brings the wet fly downstream swinging broadside to the current, giving trout the best look at it. The wet fly should not travel faster than a natural insect might swim.

Set up a pattern, casting short at first, then working each cast farther out, until you're fishing a comfortable distance for you, forty to sixty feet. This is about the maximum at which you can control the swing and mend the line. When your casts are extended to the limit of your ability to be graceful, begin taking a couple steps downstream between casts, and present your fly to any hidden lies along the length of the riffle. Mend almost constantly throughout each swing. Fish the entire riffle, if it's deep enough and slow enough that trout might hold anywhere along it.

When you come to any prime lie, slow down and fish your casts more carefully. Try to move in closer, and tend the drift of the fly so that it probes in and around the lie. Mend line so the fly hangs in water where you would expect a trout to be.

If the riffle is shallow and fast, appearing to be empty water or bordering on it, stalk only the prime lies. Fish them from upstream and off to the side. You'll often take trout from surprising places this way. They'll hold behind rocks of softball size in just inches of water at times. But you're not likely to catch large trout from this type of water. Big ones hold in prime riffle lies where the water is at least a couple feet deep.

Fishing the Surface

Fishing dry is the easiest, and often the most productive, way to explore a riffle. Dry-fly fishing in fast water is far easier than fishing upstream nymphs, and it's a little easier than fishing wet flies if you fish wets beyond the chuck-and-chance-it level.

Trout in riffles are seldom selective unless a whopping hatch is in progress, at which time you must match it. Still, a dry fly chosen for its resemblance to insects that trout eat in riffles will catch a lot more trout for you than a fly resembling nothing riffle trout have ever seen.

Because adult caddis are so active, out and bouncing about so much during the daylight hours, the best flies with which to explore the tops of riffles are impressionistic caddis patterns. I have had the most success with the light-colored Elk Hair Caddis and the darker Deer Hair Caddis, usually in sizes 12 and 14. These two dressings resemble the most common types of riffle caddisflies. They float well. The Elk Hair is very visible, even in failing light. But there are times when the brighter fly fails to draw up fish.

Whenever this happens, switch to the drab Deer Hair Caddis; trout seem willing to come up for it if they're willing to come up for anything at all.

Riffle trout won't often ignore you if you show up to fish them with Royal Wulffs, Humpies, Stimulators, or any other favorite dry-fly patterns you possess. Fish with the flies in which you have the most confidence, because riffle trout are seldom selective. The best sizes top out at 10 or 12, but I often find size 14 a good starting point and use size 16s most often on riffles that are fished more than occasionally by other anglers.

If you're exploring a riffle and don't know whether trout want a dry or a nymph, try suspending a small generic beadhead off the hook bend of your dry fly on about two feet of tippet. The first rule is to use a dry a size or two larger than the nymph, so it will float with the added weight. The second rule is to use a tippet to the nymph one size finer than that to the dry, so when you lose the nymph to a snag or feisty trout, you don't lose everything and need to replace the whole rig. When the floating half of your dry-and-dropper tandem disappears, don't stand and wonder where it went; set the hook.

When fishing a dry fly or a dry and dropper, it's best to start at the foot of the riffle, as you did with the deep nymph, and work your way upstream. Each cast should be made at an angle that is at least a little off to the side from straight upstream, in order to keep the line from sailing over the trout before the fly lights on the water above it. Work each successive cast a bit farther out into the riffle, with the drift covering a slice of water a foot or two from the first, until you have covered the water a comfortable cast upstream from you in a disciplined fashion. Then wade ten to twenty feet upstream and cover another section of the riffle. Don't stop until you have reached the very corner of the corner.

If the water is not deep enough to hold trout throughout the riffle, work your dry-fly casts over just the prime holding lies. Be sure to cover the water at the edges of the current; along any seams; over any trenches, ledges, or shelves; and above and below any obstructions.

If you fail to bring trout up from what look like prime lies, chances are the riffle does not hold them anywhere else or holds them only when insects are active. If the tactics for fishing all three levels fail to produce for you in a given riffle, it's time to give up on the riffle and drop downstream to fish the run below.

6

Finding Trout in Runs

Runs are where riffles go when all the excitement is over. In most trout stream situations, runs consist of relatively fast flowing water between three and six feet deep, with fairly even depth and current speed from the head down to the tail. But the size and depth of a run varies widely with the size of the creek, stream, or river.

Many runs, though not all, are more easily read than riffles. Because lies in runs are deeper and the current is less boisterous, territories tend to be larger and less stressful, though not necessarily more bountiful, than they are in riffles. In riffles, trout hold stations in territories and accept what comes to them. In runs, trout establish stations in territories but often use them as bases from which to go hunting, perhaps into riffles.

You're much more likely to catch an outsize trout in a run than you are in a riffle.

STRUCTURE

Runs vary from about a foot and a half deep in the smallest creeks up to eight or ten feet deep in the largest trout rivers. They tend to be deepest, in cross section, where the flow is strongest, usually in the center if the run is straight, and shallower toward each side. If a run is located on a curve in the river, it deepens toward the outside and shallows up toward the inside. Some runs, especially those that are so gentle they approach being flats, carry almost the same depth from side to side.

Many runs gradually deepen from the head to the tail. A common case would be a run that is two to four feet deep at the head, four to six feet deep toward the tailout. But there are no abrupt drops in depth, or the water would suddenly slow down, and it would become a pool rather than a run.

The head of a run usually receives a riffle, sometimes in a shelf, often in what is commonly called a "current tongue." The water is choppy at the

It's difficult to say exactly where a riffle ends and a run begins. If Jim turned and cast to his left, he'd be hitting a riffle corner. If he cast straight to his front, he'd be hitting the transition from riffle to run. But casting as he is, down and across, probably with a weighted streamer that he'll allow to achieve some depth and then fish around on the swing, he's probing the depths of a run.

It's possible that the run should be defined as all the water downstream from the shelf that cuts across the stream. In that case, what I call *riffle corners* would have to be reclassified as *run corners*. That might make more sense. The water from the shelf downstream is more consistent with the water I classify as a run than it is with the water upstream on the riffle, where it's too fast and shallow to hold any trout. Perhaps it is more logical to make the transition in definitions abrupt at that shelf, rather than grading the riffle into the top end of the run. A sophist could get me to switch my definitions quite easily.

upper end and gradually calms as it moves toward the lower end. Many riffles grade into runs, and it's not always precise where the riffle ends and the run begins, though it's usually where the gradient becomes less steep and the surface less choppy.

Runs have a tendency to be narrower at the head, spreading out a bit as they move down and deepen, though some retain the same width from top to tail. The lower end of a run sometimes spills straight into whatever kind of water comes next, usually another riffle or a sudden or gradual deepening into a pool. But most of the time there is a lifting tailout over a buildup of cobble and stones before the run drops over and down into a riffle.

The gradient of a run is not as steep as the gradient of a riffle. Runs are not so swift, though most have strong flows. Their surfaces are smoother

A spate-stream run, or as it's more commonly called, a freestone run, has a rocky structure on the bottom and holding lies more often defined by boulders and other obstructions to the current than they are by undercuts and indentations along the banks. In this case, on Pennsylvania's Stoney Creek, the root wad of the fallen tree backs up the water a bit, slowing it, and has some depth dug out in front of it and alongside it. Though trout might be sprinkled in lies anywhere in the slight depths downstream from the boulders to Masako's left, the prime lie in this photo is associated with that root wad, and all the water around it should be explored carefully with a dry fly or nymph.

than riffles, lacking the choppiness that reflects shallow water flowing swiftly over a cobbled bottom.

The slightly slower current of a run allows smaller rock, finer sand, and sometimes even sediment to drop out of the water and settle to the bottom. As a result, the predominant bottom material tends to be small rocks, pebbles, and coarse sand, rather than the uniform cobble of riffles. The bottom of a run is more compact and does not have nearly the number of spaces between rocks that are found in riffles. Overlaid on the finer bed of the bottom, in freestone runs, is a deposition of rocks and boulders that are often much larger than those found in riffles.

Anybody who has tripped over uneven stones in a pushy run can tell you about its bottom. You can nearly jog through some stable riffles, but you must be more careful when wading most runs, sometimes having to feel your way ahead with a staff.

A meadow stream run tends to have fewer easily read holding lies than one in a freestone stream. It will usually be smooth on top, perhaps grading into a flat, except that it's generally deeper and has a more defined current. This small run, on California's Lost Creek, pushes its current along the right bank. That current not only erodes the deepest depths, but also delivers whatever food trout see on this stretch of stream. The bank is visibly undercut on that side and has a few indentations. Add the bit of shade and darkness the high bank has begun to cast on that corner downstream from the trees, and you have a whole set of reasons why Masako has read it as the best place to cast.

Meadow stream runs have most of the structural characteristics of typical freestone runs. Their depths tend toward three to six feet, fairly even from side to side and end to end. They shallow toward the inside of a curve and deepen toward the outside. But they lack those bouldery bottoms. Although there might be an occasional large rock that the stream has eroded its way around, for the most part the bottom consists of pebbles, gravel, and in some cases sediment. The smooth bottom of a meadow stream run is reflected up to a glassy surface. If the source of the stream is stable, such as a steady springhead, vegetation might take root, though plant beds are more often features of slower flats than of faster runs, with their slightly steeper gradients.

The surface of a freestone run is calm where the bottom is fine but rougher where it flows over rock and boulders. If a boulder protrudes above the surface, it is easy to see exactly how it forms lies for trout, in the

pillow upstream and the slick that extends downstream for five to twenty-five or thirty feet. Wherever large but invisible obstructions break the current on the bottom, boils and slicks might quarrel on the surface. These tell you about fine lies for trout down below.

Many lies in runs are not reflected on the surface at all, and you need to search for them with your flies, usually nymphs fished along the bottom. These lies will be revealed only when you catch trout from them. Catching a trout is not the worst way to read water and discover the trout's location.

NEEDS OF TROUT

The structure of a run differs from that of a riffle, and it meets the needs of trout in different ways. Runs often hold the territories and stations of trout that move in and out of feeding riffles. Because runs offer larger territories,

they usually receive trout that outgrow riffles and move toward bigger water. If your goal is to catch larger trout, one way to achieve it is to spend more time fishing runs and less time fishing shallow riffles, though you should not neglect any prime lies in riffles, and also keep in mind that a trench, boulder, or ledge lie in a riffle might form a shelter in which a lunker can laze and from which it can shoulder its rifle and go hunting.

Shelter from Currents

The gradient of a run is not as steep as in a swift riffle, the current is not so rushing, and the need for shelter from it, though still a major factor, is not quite so critical. Trout are also less territorial as current slows, so a large and comfortable lie in a run might hold several trout, whereas most lies in riffles hold one trout per territory, though the territories might be so close

This run, on the Stillwater River in Montana, is a perfect example of a boulder garden. Its currents glance from rock to rock, some of them exposed, others showing as broken water on the surface, and many not showing at all but still holding trout. It's easy to read current lines in such water, to tell where trout find places to avoid the current and also see where food will be trotted to them. The obvious lies are in relation to boulders, upstream and down. But less obvious lies form under current seams where one boulder shoves the current aside and that current meets another shunted out of its path by another boulder.

together that you think you're fishing a pod when you catch several from one spot.

Large boulders found on the bottom of a run interrupt the flow, creating dozens of the kinds of lies that are more scattered in riffles. They also cause turbulence, which sounds bad at first, because wherever water is continually tossing around in different directions, trout cannot orient themselves and hold comfortable stations. They'll avoid setting up territories in seething water, though they'll move through it or into it to take food when it's offered there. But turbulence often creates holding lies in runs where bottom stones are not individually large enough to form them.

The word *turbulence* brings up violent connotations. But take a deep breath, clasp your nose, go down below, and pretend you're a trout trying to find a lie near the bottom of a run. Here comes the water right at you, with all its force. Hold your breath, brace yourself, and get ready to wag

Though they're not always empty water, long, broad runs that are of uniform depth and show little evidence of anything on the bottom to break the current, such as this one on Oregon's Willamette River, will have at most very scattered trout, and you'll have no sign of where to find them unless they're rising. It's often best to explore this water with a few casts to see if trout might be more abundant than you expect, then climb back into the boat and continue on your trip downstream, looking for trout in water that is more readable.

your tail with all your might. It looks as though you're going to have to swim a sprint just to stay where you are.

But a bunch of medium-size stones just upstream from you, none of them large enough to form a lie by itself, bounce the current back and forth between them. The result is a bunch of swirls and eddies that would reveal themselves only if somebody upstream from your lie along the bottom poured dye into the water. As a trout hugging the bottom, already taking advantage of the rules of friction that slow the water in a layer there, you feel small soft spots of gentle water where the currents cancel each other. You can hold there without much effort at all.

You've been a good trout. Come on up now.

The general trend of turbulence, according to hydrologists, is upward, away from the bottom that causes the turbulence. Because of both friction and turbulence, the layer of slow water formed along the bottom of a run is substantial, a zone several inches thicker than that offered in a riffle. Shelter from strong currents, therefore, is offered in a lot more places in a run, and trout find a lot more water where they can hold in comfort.

The obvious exception to this is any run without much other than bedrock or fine cobble on its bottom to break the current. Even if the flow is relatively easy, a trout will not fight it constantly. If a run has no obstructions, it will have very few trout, and rarely any of any size. Read the water in any run: If it's entirely featureless, it will usually be troutless.

Protection from Predators

Protection from predators is offered primarily by the greater depth of runs. The surface of a run is, if not a broken window like the surface of a riffle, at least a distorted window, like an old, dirty pane of glass. Overhead predators cannot see trout through it clearly unless the trout are feeding high in the water, holding up near the surface. When trout are on their stations, down near the bottom, distortion combines with the dimness of some depth to make them nearly impossible to detect.

Not many aerial predators could puncture enough water to take a trout at the depths where they commonly hold in runs anyway. But runs are happy hunting grounds for diving birds, such as mergansers, and for the sleek whiskered death that is the working end of an otter. The fish are aware of it.

Trout are alert in proportion to the shape of a run. If the run is fairly fast and its surface rough, they'll feel relatively secure. If the current is mild and the surface smooth, they'll have their senses honed for danger. You must make an assessment of the kind of water they are in and approach each accordingly. If you always err on the side of quiet and caution, you will seldom frighten your fish. You might even see an otter. If you jog right up in a shower of spray, you can't expect to catch trout, no matter what kind of water they're holding in.

Trout Foods

It's critical to recognize the difference between freestone runs and those with stabilized flows, in spring creeks and tailwaters. Freestone runs tend to have lots of lies, formed by boulders on the bottom, that are sheltered from the current. But they also tend to be less insect rich than riffles. The result is an abundance of holding lies providing the first need of trout, shelter from the current. These same trout move toward feeding lies to meet their second need. As a consequence, you'll often find trout in freestone runs only where their holding lies are associated with nearby feeding lies.

In runs with stable flows, vegetation takes root or photosynthetic growth on bottom rocks is not scoured out by annual spates and sometimes becomes luxuriant. Such vegetation forms myriads of niches for aquatic insects, crustaceans such as sow bugs and scuds, and even aquatic worms.

Spring creek and tailwater runs offer trout many prime lies, where they find shelter from currents, protection from predators, and an abundance of food.

Most mayfly nymphs in freestone runs are crawlers, some sizes 12 and 14 and blocky, but just as often the more slender types such as pale morning duns (*Ephemerella*) and mahogany duns (*Paraleptophlebia*). A few of the largest clingers, size 12 to 16, such as the Gordon quills (*Epeorus*) and pale evening duns (*Heptagenia*), also live in runs, usually migrating to the calmest edge waters for emergence. Because the surface is always smoother on runs than it is in riffles, when you get into a hatch of duns resulting from the presence of these nymphs, you'll usually find it necessary to match it with at least an approximation, in terms of size, shape, and color, in order to catch many trout. Mayfly nymphs in spring creek and tailwater runs tend toward the same pale morning duns and mahogany duns, but because they emerge on somewhat smooth waters, using that primary defense against predation known as mass emergence, you'll almost always have to match them with close imitations to fool the selective trout of such waters.

Caddis larvae in freestone runs, like mayfly nymphs, also run larger, and they tend to be cased rather than the free-living types that are most abundant in riffles. Because of their need to remain fixed on the bottom in currents that are at least pushy, if not strong, they almost all incorporate sand grains and even tiny pebbles into their cases as ballast. These caddis types include a wide variety of families, genera, and species; examples include the large size 4 to 8 fall caddis (*Dicosmoecus*) and much smaller size 18 to 20 turtle case makers (*Glossosoma*).

When caddis emerge, trout feed heavily on rising pupae. Though they're not always selective, you'll never make a mistake by getting a look at a specimen and matching it as closely as you can with a soft-hackled wet fly or a Gary LaFontaine Sparkle Pupa pattern in the same size and color. When trout rise to adults, an Elk Hair Caddis, Deer Hair Caddis, or specific imitation might be in order, depending on the smoothness of the surface of the run and the selectivity of the trout feeding in it.

Freestone runs in big rivers, especially western waters such as the Madison, Big Hole, and Deschutes, are home to the largest of the stone-flies, which are almost always the largest of the aquatic insects in any stream or river. Giant salmon fly nymphs live in the thousands in the broken water of fast runs. They are considered by many to be the most important single hatch of the year. But their importance, from the view of the trout, is not confined to their late-spring hatch period. Salmon flies and other large stoneflies have three-year life cycles. The nymphs of early year classes are out in the runs, available to trout, all year long. They are a staple in the diet, down along the bottom, throughout all the seasons. Their adults, which hang along the edges and usually deposit their eggs over riffles, are far less important than the nymphs in runs.

Spring creek and tailwater runs are not often prime water for stonefly nymphs, with their need for high oxygen levels and therefore broken water types. The exception, for some reason, is the western *Skwala*, which is important on many tailwaters in February though April.

Crayfish are common among the larger rocks of freestone runs. They are scavengers, living on both plant and animal life, whatever they can catch in their claws. They forage most when the light is dim and through the night. Trout come out to hunt them down whenever they can find them, most often at dawn and dusk.

Smaller crustaceans, scuds and aquatic sow bugs, adapt to niches in vegetation and can be abundant in spring creek and tailwater runs. You should always carry imitations of them if you fish such waters. A few San Juan Worm dressings, imitations of aquatic worms, will serve you well on waters where the naturals are abundant. These same flies will also serve as searching nymphs on almost any moving water that contains trout.

Temperature and Oxygen

Runs offer excellent temperature and oxygen regimes in all but the most stressful conditions. A stream has to be on the brink of disaster before trout will find any benefit in moving out of runs and into highly oxygenated riffles. Such conditions do occur. In some damaged or desert watersheds, they happen almost annually. But most of the time, trout maintain their stations in runs even in hot weather.

When conditions do become extreme, many of the springheads that cool a stream are found in runs. Wherever a side stream makes its entrance in a run, you can expect to find fish holding in the cooler water just downstream.

TYPES OF RUNS

There are four basic types of runs: barren runs, feeding runs, holding runs, and prime runs.

Barren runs have little life in them, at least of the kind that makes them attractive to trout. The most common kind of run that lacks life is one that has a bottom with little rock or rubble or with stones so small and uniform that they form no lies where trout can avoid the current.

A run with a silt or sand bottom will not have a great deal of insect life unless it has lots of rooted vegetation. But freestone runs are subject to scour, and vegetation does not get a chance to take root. The only kind of silt- or sand-bottomed run that is rich is one with rooted aquatic vegetation, but any run that is slow enough, shallow enough, and safe enough from winter scour to allow the growth of vegetation suffers in this book from getting classified as a *flat* and is discussed in chapter 8.

Another kind of barren run is the long, slow, and turbid kind found in the lower reaches of major rivers. These might have rocky bottoms, but in most cases the rocks are so covered with silt that there are few places left for insects to live. The water is often opaque; sunlight does not penetrate to the bottom. Photosynthesis could not take place if a bit of algal growth could find a clean stone to stand on.

Feeding runs are rare. You won't encounter them on many streams, but I have fished a few. Trout do not hold on them, for lack of lies where they can avoid the constant current, but they move up into them whenever a hatch is on. These runs are shallow, two feet deep or less, but with less gradient than riffles and slightly smoother surfaces. Some are fast, bordering on riffles. Others are slow, bordering on flats. Fish nose up into feeding runs whenever a food form is active and conning trout into risking aerial predation. When the source of food disappears, trout will move out of these kinds of runs into water where they can find protection from predators.

Holding runs offer trout shelter from currents and protection from predators, but they are not rich in opportunities to feed. This does not mean that trout merely hold in them and nap until they move out into a feeding run to stock up. A holding run delivers some food down its currents, and trout hold their stations, defending their territories from intruders, feeding on what washes their way. They do not leave their lies unless enticed out of them by an abundance of food in some other kind of water nearby.

Most holding runs are two to four feet in depth and fairly homogeneous in their bottom structure. If anything defines them, it's a lack of rooted or attached vegetation and a scarcity of spaces between rocks in which insects can hide and grow. Often these types of runs have sandy or slightly silted bottoms, with stones embedded in the finer substrate. Rocks entrenched this way deflect the current, forming lies. But bottoms like these don't produce much trout food.

Holding runs always hold trout. But the size of the trout, regardless of the size of the territory, is dictated by the amount of food available nearby. As Charlie Brooks said in *The Trout and the Stream*, "It is true that the larger the fish the larger volume of water he requires, but first he requires sufficient *food* to make him large." Most trout found in holding runs will be small to average in size. Large trout will not remain in such lies for long periods of time unless there is a dependable and prolific source of food very near them.

Types of Runs

Feeding runs border on empty water and most of the time are a waste of fishing time unless you see trout active in them. This one, on Lost Creek in California, is too shallow and exposed, with too few obstructions on the bottom to break the current, even though it's not very brisk. But a hatch of insects might attract trout seemingly out of nowhere, though most likely out of the deeper water upstream, in the shade of the trees.

Often trout that appear out of nowhere were right there all the time, simply tucked into almost invisible lies along the bottom and very reluctant to feed because of their exposure to danger. If you fish delicately enough, with a small, white yarn indicator and size 16 or smaller beadhead nymph, you might find trout that seemed absent and discover that what looks like empty water or a feeding lie, occupied only during a hatch, is actually a holding run and has trout in it all the time.

continued on next page

Barren runs, such as the upper one on the Big Hole River in Montana, are shallow and featureless. They have unbroken currents with few places for trout to take shelter, leaving them exposed to overhead predation. In general, they are the kind of water you just want to get through in the boat so you can fish water downstream that has more potential to hold trout. The water under the boat in this photo looks barren, and it proved to be when I fished it. But the anglers were wise to keep their flies in the water, or on it if they were fishing dry. Sometimes water that appears to be barren provides surprises.

The water downstream from the boat, where the run tips a bit, shows signs of boulders on the bottom and looks far more productive. In fact, I fished it from shore before the boat arrived and hooked a couple nice trout, one a sixteen-inch rainbow. This put the fish down and caused the other anglers to go through it without hooking anything, which probably led them to think it was an extension of the barren water upstream. If you're wading, as I was, then the ability to recognize the rumpled water as productive, and the water upstream from it as likely to be empty, allows you to high-grade the water and beat boaters to trout.

Types of Runs continued

This holding run, on the Big Thompson River in Colorado, drops off to fair depths along the bank across from Terri Yamagishi. It also has some slightly seething currents on its surface, reflections of at least a few boulders on the bottom in the central part of the run. These are all signs that trout hold in the run at all times and can be caught, as Terri proves, if you adjust your tackle and tactics to the prevailing conditions.

This run on the Yakima River, fished by Skip Morris, is prime, though its trout might not always be easily caught. It has depth and considerable current to deliver food, and its rumpled surface proves that it has plenty of boulders on the bottom. Trout in such water will find comfort and abundant food and will reside there at all times, rarely moving out to feed anywhere else. The problem is getting a fly down deep enough, and moving slowly enough, to entice them to take it. But most prime lies offer similar problems. If they were easy, they would not often be prime.

Prime runs are what we think about when we clasp our hands behind our heads, lean back in our winter chairs, and daydream about trout water. These runs combine the aspects of feeding and holding runs: They contain lots of prime lies that fulfill all the needs of trout in one spot. They have sprinklings of boulders to give shelter from currents. They're usually three to six feet deep, providing plenty of protection from predators. Most important, they have bottoms of clean rocks in mixed sizes: lots of living spaces for aquatic insects and even a thin but rich layer of algal growth to feed them so they can grow and be happy and feed the trout.

The broken bottom features in prime runs also provide shelter for other organisms: crayfish, baitfish, and even some sinuous and nasty leeches. Big trout thrive on these big bites and find that they don't ever have to leave their territories to get any of their basic needs met.

HOLDING LIES

Reading runs to locate prime lies is more critical than it is in riffles. Trout are less scattered; they hold where combinations of shelter and security and food are as good as they can find according to their size. Where the water is prime, trout establish stations that are fairly close together, though each station gives a window onto a separate territory.

The territorial imperative declines as water speed slows. In very slow runs, trout almost bunch up in pods. This tendency to gather in the best water is the reason you sometimes fish through most of a run without getting a bump, then suddenly hook trout after trout in a pocket of limited extent. It happened to me once on a run in a river in Montana.

The river was medium-size, about two casts wide. It alternated between long riffles and short runs. I fished a riffle with a wet fly. The water was shallow, sunlit, and rich but had few holding lies, and I coaxed no trout to my fly. I moved down to the run below but didn't change from the downstream wet-fly strategy I'd tried fruitlessly upstream.

The head of the run, where the riffle broke into it, looked good. But no trout moved to my fly there. I fished around a few boulders, where the water promised trout, but the promises were not fulfilled. Finally I waded into the center of the run, which was only about three feet deep, and began casting alternately to both banks. I worked along for about fifty feet this way without a tap. Then my wet fly began to make an arc that carried it toward the edge of a circle of shade dropped from a pine tree leaning precariously from the left bank, about to topple into the patch of shade it cast.

Trout Lies in Shade

When the sun is out, and all other things are about equal, look for trout in shade. If all other things are not equal, such as if the water is deeper under a tree or a current line enters with the probability of delivering food, trout might be gathered there.

I'm never sure whether trout in the shade have moved here from other parts of the stream or have simply been inactive in bright sunshine but suddenly become active when shade covers them. I suspect it's a combination of the two: Some trout move from exposed lies into shade, and other trout that were in sunshine become willing to take flies when shade moves over them.

The shallower and clearer the water, such as this run on upper Hat Creek in California, the more likely trout will be dour in sunshine, feeders in shade.

Another reason trout hang out and are active in the shade of trees is the great abundance of insects that often gather in the leaves and then fall to the water.

Many of these are golden stonefly adults, making necessary preparations for the adventure of flying out over water to lay their fertilized eggs, but caddisflies and all sorts of terrestrial insects prowl about in trees as well and make mistakes that propel them onto the water. Trout know about all this, and shade is far from the only benefit they get from crouching under the overhanging branches of a tree.

When the fly skirted the darkness, a trout came out and rapped it. It was a brown, about fourteen inches long. I played it, netted it, released it. I cast again to the same place, and again a brown went *whap*.

The length of run flowing through the shadow of that small tree was no more than twenty feet. In that short distance, I took fifteen trout, all between twelve and sixteen inches long. Some were browns, some were rainbows. All were fine, and all held in the shade. When I fished through it and then continued beyond it, I got no more strikes.

On an impulse, I decided to wade back up through that shade to shore, to get out and take a rest in it. But just outside the shadows, the bottom began to get away from me. I tried to wade through, but before I reached the area that had held the trout, the water was trying to tickle my armpits. I had to back out of there and go down and around the shadow to reach shore.

Since I didn't recognize that deep pocket until I almost fell into it, since I didn't go right to it and fish it immediately upon arriving at the run, what business do I have writing a book about reading trout water? It's a good question. The only answer I have is that I did discern the most likely holding areas in the run and did fish them one at a time until I stumbled onto the one that held trout, which I should have recognized at once as separate because of its shade and depth.

By learning to recognize high-percentage lies, and by fishing them one at a time, we eventually stumble onto a piece of water that produces trout. Viewed in that regard, it's not really a stumble. *Movement is always a major element of a successful fly-fishing strategy.*

Though it's a minor digression to say it here, the same thing is true of tactics. The angler who tries this, then that, and finally the other thing will, if he can keep things to a level just under a frenzy, have fun and eventually find a tactic that takes trout. To an observer, it might look like an accident. It's not: *Change is another of the major elements of a successful fly-fishing strategy.*

Head of a Run

A riffle feeding into a run delivers an almost constant supply of drift. The head of a run, with its corner pockets filled like the cheeks of a chipmunk, is second in productivity only to the corners of riffles. Sometimes it is not even second.

The richness of the riffle upstream is the direct cause of the productivity of the head of a run. You'll do well to glance at the riffle that feeds into any run; it's a part of assessing the run's potential. A brief riffle will produce insects, but not as many as a long one where the water bounces down

Not all runs have riffles at their heads, but the most productive water of a run is usually at the upper end, where the currents enter. Food is most abundant there. The entering current breaks up the surface and provides more protection from overhead predation. Any slight corners or other obvious holding lies become prime and hold trout at all times. If there is such a thing as a typical scenario for fishing a run, it might be that you enter at its upper end, catch several trout of small to medium size by covering all the likely water there, then begin the long process of fishing down the rest of the less featured part of the run, where the water is generally deeper and the holding lies are much less likely to be marked and therefore readable. In that long reach, you catch one to at most three trout that are commonly much larger.

whitely, so thin and fast no trout could hold in it to pick over the current's offerings before they get delivered to the run below.

The head's productivity is also based on its potential as holding water. If it forms as a chute with little to break the current, there will be few places for trout to hold, and they will move into the head of the run as they do into a feeding riffle: only when there is such an abundance of food that it makes fighting the current worth the energy expended. If the head of the run has a relatively slow current, with gentle eddies off to the sides, trout will establish territories and stick to their stations.

If the head of the run shelves off immediately into fairly deep water, from four to six feet, it's very likely to be a prime lie. This kind of lie located immediately downstream from a long, rich riffle can hold the largest trout in the stream or river. Trout find shelter from currents, protection from predators, and an abundant supply of feed and often stack up there. A small territory is sufficient. It's the kind of lie where you can sometimes take your limit of trout without moving your feet, if you are interested in taking a limit of trout.

Visible Obstructions

The next most likely place to find trout in a run is around any visible obstruction along its length. The obstruction that first leaps to mind is a midstream boulder. It attracts trout for the same reason it would in a riffle: Anything that breaks the current provides shelter from it and creates a likely holding lie. If the water surrounding a boulder is deep enough to provide protection from predators, the lie behind it creeps quickly toward the prime class.

A large boulder visible in the current provides more than a single holding lie. The pillow of water just upstream is often the most productive water, for the biggest trout. It forms a comfortable station, with an unblocked window on the current. It's also the easiest water to fish around a boulder. A dry fly set to the water two or three feet upstream from the boulder, with lots of slack tossed into the line and leader, will drift down to the lie naturally. A weighted nymph cast five to fifteen feet upstream, depending on the current's strength and depth, will sink down and tumble right into the pillow, before it's drawn off to the side by the current that breaks around the boulder.

Each side of a midstream boulder has slight delta-winged eddies. The current gets busy like a whisk broom and hollows out small depressions at the base of each wing, tucked right down under the edge of the boulder. The current delivers bits of food that swirl into these eddies. Trout hold there, tight to the boulder and the bottom, especially if the boulder casts a bit of shade. The problem is the difficulty in delivering a fly to these wing lies. The tug of the line, cast over conflicting currents between your casting position and the boulder, draws the fly out almost every time. About the only way to defeat this is to wade very close, lob a heavily weighted bomb, and fish it right underneath your rod tip.

The storied big-fish lie in almost all fishing fables is the eddied water behind a big boulder, in a deep and dark run. The reason is simple: This kind of water holds lots of trout, sometimes very big ones. The water immediately behind a visible obstruction is almost still. A few feet to as much as thirty feet downstream, water that has separated and swung wide around each side of the boulder forms a seam until the currents get back together.

Wherever you find a boulder in a run, probe it carefully, no matter what sort of fly you're using. Fish both its upstream and downstream ends, and if possible, run your fly over or through the water tucked in tight to both sides of it. The closer you can wade to it, shortening your cast, the better you'll be able to fish it.

Trout hold close in behind a boulder. They also hold in the eddied water downstream a few to several feet from it, but *right along the bottom*, beneath the confused water on the surface. If the water is shallow, they'll drive to the top for a dry fly. But they'll hold along the bottom.

I have my best luck fishing boulder lies in runs that are relatively shallow, two to four feet deep, where trout are willing to rise for dry flies. It's difficult to get a decent drift in the eddy behind a boulder. I try to move in close and keep the dry bouncing on the water immediately behind it or down one of the two seams behind it, before the split currents meet. A trout gets a bit of a chance to see the fly before drag sets in and will usually slash up quickly if it's going to rise at all.

Your line wants to pull the dry fly out to the sides or downstream, creating almost immediate drag. To prevent this, hold your rod as high as you can, lifting the line over the conflicting currents below the boulder. This extends the drag-free drift. The closer you can wade, the shorter you cast, and the more line you can lift off the water, the better your chance of keeping your fly dancing where you want it until it goads a strike.

The same problems apply when trying to get a nymph down to trout holding behind boulders in runs. The nymph has to plummet like a stone if it's to get down before it gets washed out of the potential holding water. It helps to use very heavily weighted flies or pinch two or three heavy split shot to the leader between a pair of nymphs. It also helps to wade as close as you can, cast short, and hold your rod high, the same as with a dry fly. You'll have a problem detecting all the takes to nymphs, even with a strike indicator, because the confusing currents tug the indicator in all directions.

Always fish out the eddied water for a few feet downstream from the boulder, after you have probed the water immediately below it. Fish either on the surface, with a dry fly, or right on the bottom, with one or two weighted nymphs. Boulder lies offer little in the way of mid-depth fishing.

Other kinds of visible obstructions should be fished like midstream boulders. A log or root wad lodged in a run offers the same kind of cover a boulder provides. Such lies are nastier to cast into, because you're more likely to get snagged. But trout will be down under the obstruction, out of the current, and you'll have to do your best to show your dry fly or drift your nymphs to them without getting hung in the wood.

Generally, the larger the obstruction in a run, the larger the trout it will shelter. But the size of the object must be in proportion to the depth of the run: A two-foot-deep run with a boulder the size of a Buick protruding out of it will not hold the same size trout as a three-foot-deep run obstructed by a boulder the size of a beach ball.

Look for large trout around visible lies in runs wherever the water looks as if it might provide for their needs. In many ways, and with lots of experience, reading all trout water boils down to a formula as simple as that.

Invisible Obstructions

Boils on the surface reflect boulders below. These lies are as prime as visible boulders, perhaps more so because their depth adds concealment and protects trout more thoroughly from predation. The first thing to do with an invisible boulder is figure out exactly where it rests on the bottom. You want to fish the boulder, not the boil.

A boulder will be from a foot to about five feet upstream from its boil, depending on water depth and speed of the current. Once you have placed the boulder precisely, spend a moment imagining it in your mind. Picture it with its pillow in front of it, its little winged eddies out to the side, its eddied water close downstream. Put trout in all these places. Once you have envisioned it in this way, fish it exactly as if it protruded above the water and you could see every feature the current created.

Boulder Gardens

A few of the most productive runs are what summer steelhead anglers refer to as "boulder gardens." These are studded by boulders, some of them large enough to protrude from the surface, more of them smaller and not breaking the top. Those that you cannot see sometimes show up as boils or disturbances on the surface; most often they lurk down there, invisible, to trip you up while you're wading. Boulder gardens can be treacherous, especially those in which the current is forceful. You must wade them with extreme caution.

Boulder gardens send the current bouncing back and forth, impede its progress, slow it down, split it, and cause it to rejoin in current seams. They provide trout numerous comfortable lies where they are sheltered from currents and, in all but the shallowest runs, are protected from predators. Because the bottom is cobbled, though usually with larger rather than smaller rocks, insects find plenty of places to live, and trout are often fed very well without leaving their stations. All those boulders, visible and invisible, add up to an exceptional number of prime lies.

In depth, boulder gardens vary from just a couple feet on small streams to five or six feet on large rivers. You can imagine the kind of wading you'll get into where boulders are big enough to break the surface in water deeper than three or four feet. It's best just to stay out of it and fish it from the bank if you can. The most comfortable boulder gardens, for the angler, are the shallowest, two to three feet deep. Because photosynthetic growth is greatest where the depth is least, these are also very productive for insects, and therefore trout.

Deeper boulder gardens are also productive, though some of them are better as holding lies from which trout can venture out to search for better feeding pastures. Such boulder gardens are also excellent holding water for passing migratory fish. If you enjoy the sullen, head-shaking takes and sudden eruptions from the water of those outsize trout called summer steelhead, you'll spend an overabundance of time wading to your wader tops in difficult boulder gardens.

Not all boulder gardens have protruding boulders to reveal their presence. In many of the best of them, all obstructions to the current are subsurface. You could easily walk by such water, declaring it empty, if you failed to notice the signs sent to the surface, sometimes feebly, by those submerged boulders. I recently walked the banks of a little-known steelhead river with Bob Wolfe, from Pendleton, Oregon, and Tilda Runner, president of the Oregon Federation of Flyfishers Council. Bob force-marched us what seemed like several miles downstream, passing what appeared to be some excellent, easily read steelhead water. But he had his

Some boulder gardens send little sign up to the surface. The only hints that this water, where Bob Wolfe and Tilda Runner wade the little-known river, might have a bouldered bottom are the slight seething of the currents and the scattered boulders upstream. When you step into it, you have no doubt about it—it's treacherous. All those obstructed currents on the bottom form perfect lies for trout, in this case the summer steelhead version of the rainbow.

mind, if not yet his eye, on a bit of water that he'd learned held fish from the best teacher of reading the water—catching fish in a certain spot often in his past.

When we arrived at it, it was as featureless as much of the water we'd passed. I asked Bob what defined it as good water. "Two things," he answered. "First, you can see the lines of conflicting currents on top, down that centerline." He pointed at some almost invisible tendrils in the currents with his rod. They disturbed the surface slightly for a couple hundred feet. The surface of the run upstream and down was smooth. "Those lines reflect boulders on the bottom," Bob said.

"Second," he went on, "you can see jumbles of boulders on the far bank. They came down off that hillside. They don't stop at the bank." Indeed, a section of the bank was bouldered. The bank upstream and down, where the surface of the water was smooth, lacked boulders. "Wherever you see boulders on the bank," Bob said, "you'll find boulders on the bottom. That's where you'll find steelhead." The thread I noticed running through all of Bob's conversation was "you see." The things he pointed out with his rod tip were there to be seen, but not the things you'd normally notice. That was, as well, an excellent lesson in reading trout water.

Tilda and I stepped into the run at the head of those conflicting currents. The wading, as predicted by all those boulders on the bank, was dangerous. Because she was armed with a Spey rod and I was not, Tilda was able to wade fifteen feet closer to the near bank than I, and at the same time cast fifteen feet closer to the far bank. I had to back out twice and take photos of her holding steelhead before I was able to finally hook one myself.

Bob caught a couple later in the day. That boulder garden, though almost unreadable on the surface, was full of summer steelhead.

Ledges, Trenches, and Shelves

Runs tend to have a lot more shelving water than do cobbled riffles, because they work their way to base rock more often. In some streams and rivers, ledges, trenches, and shelves are the primary features, and you don't hook trout until you run a fly deep through them.

You can usually spot a trench by the slightly darker water on the surface directly above it. But it would be dishonest to predict that all of them can be noticed this way. I've fished excellent runs in which the holding water could be discovered only by wading and feeling for sudden depths. It's best to do this kind of exploration by probing with a wading staff, unless you don't mind buffaloing downstream in pursuit of a floating hat.

Anybody who knows how to read water can approach a run he's never fished before, with trenches that show on the surface, and figure out how to find trout. But rivers with runs that have invisible ledges, trenches, and shelves are the kinds of rivers that produce best for somebody on the home team. You have to know these waters to fish them at their best.

Any book on reading water would be missing some sentences if it didn't point out the advantages of fishing water that you've fished over and over. You learn where trout hold by having a history of remembering where you've caught them before. You know where some hidden lies are by having a history of stepping into them. By owning a home river, or being owned by one, you not only learn its lies, but you learn more about the characteristics of *all* lies, in all trout water. Everything you learn on your home river will go with you to any river you ever fish.

Back to ledges, trenches, and shelves.

If a trench is etched into the bottom parallel to the flow, you should fish carefully down the length of it. Work a dry fly over its surface. Cast wets so they swim across the top of it, a couple feet farther downstream on each

The Advantage of Home Water

This isn't the prettiest kind of water to fish, nor is it the easiest to read. If anything, it's a testament to fishing the same bit of water consistently, learning it so that even when it's out from glacial melt, you have an idea where you might find its trout. In this particular case, I read the water on an earlier trip by attempting to wade across to that overhanging branch just behind Masako and stepped into a hole that flushed water into my waders. I was able to backpedal out of it. Next time back on this river, I added a couple split shot to Masako's nymph rig, slid the indicator two feet higher up her leader, and told her right where to cast to dangle her nymphs into the trench in which I'd almost enjoyed a swim. It took a lot of casts, but finally her indicator dipped down, and the trout she took out of there made my unusual way to read water worthwhile—though it does seem that I deserved to catch the trout.

The overhanging trees that mark the trench into which I stepped, and from which Masako hooked this trout, are far upstream in the background. She didn't hook any other trout on that short trip, but she came back feeling that this one made the trip worthwhile.

successive cast. Drop nymphs so they sink to the bottom upstream from the trench, reach the lip, and tumble into it to sweep the bottom of it.

Where a shelf is cut across the stream, the result is a sudden deepening that both breaks the flow and creates a vertical eddy in the current. Trout hang in the slow water just downstream from the shelf. They enjoy the depth, the ease, and the churning of insects that eases over the lip and swirls into the eddy. In this kind of situation, you should present dry flies so they ride the currents just below the shelf, bouncing on top of the eddy and riding the current downstream from it. Cast nymphs upstream from the shelf, then allow them to wash down with the current, free-drift, and drop over the shelf just as a dislodged natural nymph or larva would appear to the trout. The action of the down-driving current, plunging over the shelf, will drive your nymph deep.

A ledge, especially if it's overhanging, forms a perfect lie for a trout, sheltered from the passing current and protected from overhead predation. If it lacks anything, it will be feed, especially if the ledge is cut into a base-rock bottom. It's difficult for ledges to form in gravel, so they're almost always a function of a streambed eroded to base rock and therefore are usually holding lies rather than prime lies. If you're fishing a stream, especially a small one, with numerous ledges, always watch which side of the ledge throws shade. Trout holding in shallow ledge water will shift positions to follow shade.

Any deepening in the water of a run, whether it's caused by a shelving trench, a depression etched out in a gravel bottom, or a natural and gradual narrowing and deepening of the river, should be considered potential holding water. Trout take stations in all such water, holding on territories or moving out to forage if there is not a sufficient amount of feed to fuel them. These depths are revealed in some cases by the darkness of the water, in others by a slick on the surface. But often they are not revealed at all, and you have to know your river to know about them.

Whatever clue you read, fish ledges, trenches, and shelves carefully, because they hold some of the nicest trout any stream offers its anglers.

Current Seams

The joining of any conflicting currents provides a potential feeding lie in a run, just as it does in a riffle. Some seams in runs are easier to spot, where the surface is not as choppy and a couple prominent currents meet in obvious conflict. But other seams are difficult to spot. Two gentle and slow currents coming together meld without much disturbance, and you have to look carefully to see the line where they meet.

You'll rarely see the joining of two currents so obvious as it is in this run on the Big Hole River in Montana. These boaters are wise to be floating in the faster current, delivering their flies to the edge where the two currents meet. If they were in the slower water, they'd be eddied, and their flies would be moving faster than the boat. As it is, they can place their flies on or in the faster flow, very near the edge of the slower flow, and trout holding on either side of the seam will see what they've come to expect: something being delivered downstream to them on the brisker water. In the normal scheme of seams, trout will hang on the soft side and dash out to feed on drift in the fast side. Whether you're afloat or wading, always try to put yourself in a position to fish both sides of a seam.

One way to recognize the convergence of two currents is to notice the types of things that divide them in the first place. It then becomes obvious where they will rejoin downstream. One clear example is a protruding boulder. Another example is an island. Wherever an island splits the flow, a triangle of tranquil water forms at its downstream point. Some of these are too still to be fishable; they deliver no food. In others, the currents rejoin with too much turbulence—too many jumbled currents for trout to orient and hold stations—and these also are not productive. But many islands split riffles or runs or have a riffle on one side and a run on the other. Where the currents of these come together downstream, trout find all their needs met in one prime place. The apex of the triangle of quiet water points straight downstream to the seam where the split currents get together again. This seam gathers insects and drift from both currents and causes considerable fallout of feed all along its length.

Drift your dry flies down the length and along both sides of such a seam. If you're fishing wet flies, cast them so they swing across the seam. Nymphs should be tumbled the length of the seam, down on the bottom, where fish feed on the fallout. This requires that you be able to wade downstream from the island, turn, and make your casts upstream along the seam, working your way back toward the island. This works best on small to medium-size trout streams. Don't try it on big and brutal rivers like the Deschutes.

Other seams occur in runs wherever currents of two different speeds come together. The two speeds can be caused by a riffle that delivers the water in two or three tendrils, split by boulders or gravel bars. A seam can also be caused by water of different depths flowing alongside each other the length of a run. The water over each depth flows at a different speed. Where they meet, they form a seam.

These are the hardest kinds of seams to read, but they do have their indications. The surface of the water on one side or the other might be a bit rougher. An eddy line, barely perceptible, might swirl downstream for a few feet before the currents have commingled enough to average their speeds. If the reason for the difference in speeds is a difference in depth, a darkening might show on one side. Such seams are doubly desirable, because they not only gather feed from both currents, but also offer the advantage of protection from predators beneath the deeper water on one side of the seam. Fish these seams like any others, along the centerline and both sides, but get your nymphs down to the bottom of the deepest water if trout refuse to come up for drys.

The most common kinds of seams are found in boulder gardens. The current splits and rejoins. It goes fast this way and slow that way. The speeds must meld. Several boulders in random placement or staggered rows bounce the currents around and cause several seams. These seams are all potential holding lies, and you should fish each one carefully.

Foam Lines

Broken water in cascades, rapids, and riffles forms bubbles on the surface. These bubbles float, gather as white foam, and are delivered downstream. Riffles lack foam lines because they whisk it away too fast, break it up too much, keep bubbles from gathering together as foam. Wherever the water slows, foam lines begin to form, and these lines become indicators of the presence of trout food, for a couple reasons.

Foam lines are the surest signs of seams, whether you find them descending from riffles, as in this case on Montana's Jefferson River, or on any other kind of water. Insects and other trout food forms do not necessarily get trapped in foam lines, but they do get caught up and delivered down the seams and current lines that foam delineates. No matter how bold, as in this photo, or how subtle a foam line might be, always consider it a prime line of drift for whatever type of fly you are fishing. Foam, in terms of reading trout streams, is the same as chalk in the hands of a college professor, though it might make its point even more clearly: Trout are here!

First, bubbles have the approximate density of insects, especially light emergers and adults trapped in or on the surface. Wherever currents form foam lines, they also form drift lines for trout food. Second, foam gathers where the currents ease off but is whisked away where they are brisk. Therefore, a foam line indicates water that is at least a bit gentler, surrounded by water that lacks foam and thus is a bit more swift. Mayflies, caddisflies, and midges all look for these softer bits of water in which to emerge, because they're more likely to survive there . . . more likely to survive the currents, though less likely to survive predation by trout.

Don't make the mistake of thinking that insects and trout pull down their baseball cap brims to block the sun, shade their eyes, and go scouting for foam lines in which to emerge or feed. They don't. It's doubtful either insects or trout ever notice the foam. But the insects do notice those softer currents and emerge in them, and the trout notice those gatherings of food and line up to prey on it. The foam is an indicator to you that trout might gather under it, either to feed selectively on the surface, if a hatch is happening, or to lie and feed opportunistically beneath those marked currents, if no insects are on top.

Wherever you see a foam line in a run, suspect that trout will be lined up the length of it. Foam lines are most often functions of runs, because runs are generally the first slow water in which the bubbles can gather as foam downstream from cascades, rapids, and riffles, forming lines that predict trout lies to you.

Lacking any other clues to fishing a run, always fish in relation to any foam line. Such a line will often be your only clue to a subtle seam, because the melding of two currents delivers bubbles and bits of foam from two directions, and the natural affinity of bubbles for each other causes the line to form where the two currents converge. A foam line can draw the outlines of a seam perfectly for you, even if you're unable to see the seam.

Foam is delivered downstream from the brisk water of a riffle into the slower water of a run and coalesces where the main current calms and flattens, thus revealing the deepest central current of the run. In the upper end of a run, such a foam line predicts the most likely lies for trout, because it shows not only where the most food is delivered out of the riffle and into the run, but also where the water is deepest and therefore offers trout the most protection from predators.

In almost all cases where you see a foam line, and not just in runs, given no other evidence of lies, that's where you should cast. In cases where other evidence is present, such as depth, a current seam, or feed lines, it will usually be accompanied by the presence of foam, and you should know to fish in that spot or along that line of revealed drift.

It's not uncommon to find trout feeding almost invisibly, with sipping rises, in foam lines, because insects in foam lines are often small and are almost always trapped in the surface film. Trout know they are unable to escape and feed on them with little of the violence that would reveal their rises to you. If you first register the foam, then watch it carefully for rises, with binoculars if necessary, you'll find feeding trout that you otherwise might not notice. It's one of the most accurate ways to read a run or any other kind of trout water: Look for foam, then look for trout taking insects with the tiniest rises. Once you've found them, you know the precise address of the trout. By watching the foam even more closely or suspending an aquarium net in it, you can also determine what the trout are taking and select an appropriate imitation.

It's surprising how often trout taking with such tiny rises turn out to be big ones.

Current Tongues

The current of whatever water type lies upstream from a run—usually a riffle, less often a cascade or rapids—is delivered downstream into it. The current enters the run in the form of a narrowing and slowing line called a current tongue. If the stream or river is of the foothill or lowland variety, with a riffle-run-pool structure, then the pool often breaks over into a short riffle, and that riffle enters the run downstream as a current tongue.

Current tongues define trout lies clearly, because they carry insect life from the productive riffle upstream into the slower, and therefore more amenable, water in the run downstream. Trout hold in the slower currents and accept what is delivered by the faster currents. As in riffle corners, trout in runs hold as far upstream as they can find comfort from currents and protection from predators, in order to be first to the feed.

Current tongues entering runs define holding lies clearly for a second reason: They are the strongest force in the run and erode the deepest depths in whatever forms the bottom. An entering current always has the deepest water directly beneath it. In a creek or small stream, this depth might provide the only protection from overhead predation. Trout will find an obstruction to the current—a big stone or small boulder—and tuck in behind it. If the water is deep enough to provide protection from predators off to the sides, trout will be more likely to line up alongside the thrust of the main current but close enough to dart into it to feed.

The main current entering a run might be central, running down the middle, or pushed to one side or the other, depending on the shape of the banks. Often the depths of a run are pushed to the outside of a bend, with a gravel bar left on the inside. In a small mountain stream, the deepest current of a run might be pushed up against a cliff on the outside of a curve. In a more typical trout stream or river, the deep current skirts the outside bank of a bend pool. Wherever you find that more forceful current in a run, it defines the line of the greatest depth and the most food and therefore predicts where you are likely to find trout in the run.

Fish the main current, whether with drys floating on top or nymphs searching along the bottom, if the water is two to four feet deep. Fish the seams on both sides of the main current tongue as well, if the water has enough depth to provide trout protection from predators. If the water is deep and the main current strong, focus your casts along the seams between the fast current in the center of the main flow and the slower currents to the edge of it. Trout will likely find restful lies there and are less apt to be in the main force of the flow unless the run has obstructions along the bottom.

Tailouts are great places for large trout of all sorts—browns and rainbows and cutts and even occasional brookies—to move out at dusk to hunt for bait-fish such as sculpins. The water shown here on Washington's Yakima River, fished by Rick Hafele out of Yakima River Fly Shop guide Guy Drew's beautiful wooden boat, might be productive in low light, but it's not likely to produce anything in bright sunshine unless lots of insects get gathered by those currents that are about to turn from a run into a riffle.

Tailout

Some runs make an exit by slowing and lifting over a buildup of gravel before entering a riffle downstream. The bottom is almost always uniform, without boulders or other obstructions to the gathering current. The water is shallow, lacking substantial protection from predators. But this transitional water is usually rich in aquatic insect life. The tailout of a run is an excellent feeding lie. Trout materialize out of nowhere to feed hungrily whenever a hatch begins.

A tailout is difficult water to fish. The surface is usually glassy, the current is too swift to wade easily, and the trout are spooky. You won't be able to get as close to trout as you would like to. You can't get good drifts with medium to long casts, which are required because you're unable to wade close. Trout usually arrive on the tailout at the invitation of a particular species of insect, so you're forced to match the hatch in order to attract the interest of the trout. The currents of a tailout accelerate as it narrows, becoming faster downstream than they are upstream. If you try to fish upstream with dry flies, your line lies on faster currents than those on which your fly lands. The line tightens and you get drag but have no way to compensate for it.

If you want to fish dry flies on tailouts, you're almost forced to fish from upstream, placing your casts downstream or at an angle across currents that are all the same speed. If you insert wiggle into your cast, and therefore slack into your line, the current will not cause drag until the slack is taken out. You get at least a bit of a drift. If you can feed slack line into the downstream drift, you'll get a longer drag-free float. It's a bit like fishing over rising trout on a slow flat, except that everything in a tailout is accelerated, making the downstream presentation much more difficult.

I've had my best luck fishing big and pushy tailouts with wet flies that approximate the size and color of whatever natural is hatching. It's easier to present a wet in a natural manner than a dry. Swing the wet slowly across a tailout, mending line to slow the fly.

Dry-fly fishing can be superb on tailouts if the water is shallow and the current gentle, not a common case. Stalk trout on such water from upstream or off to the side, pinpoint their rises, and present your flies right down their feeding lanes. Because the water is thin and usually clear, avoid lining the fish. Approach from upstream, cast down at a quartering angle, and feed slack to your fly so it drifts downstream without apparent attachment to your leader, your line, or you.

FISHING STRATEGIES

Evident lies in a run, from the current tongue at its head to the gathering water of its tailout, are all places to focus your fishing. Each of these different kinds of holding water has the potential to turn up concentrations of trout. Knowing how to read the various water types will help you locate such a concentration more quickly and thus spend your time fishing water where you're most likely to catch trout.

It's also possible that trout might hold on invisible lies in a run, scattered throughout as they often are in riffles, in water where you're unable to pinpoint them or assess their most likely locations. When you can't isolate specific holding lies, it works best to set up a disciplined casting pattern, covering most of the potential holding water in the run.

Trout can be moody in runs. In riffles, they tend always to be feeding actively or at least willing to accept something that passes in the drift, on the top or the bottom. But in runs, trout activity is more closely tied to the

Sometimes big runs are unfeatured, and you're forced to use your fly, or a pair of them, as your eyes, to read the water by finding trout for you. Cast long and explore the run, as Skip Morris is doing here.

activity of what they eat. If aquatic insects are on the move, trout will be moving too. If nothing is happening, trout tend to hold on their stations, feeding only when an opportunity presents itself. They'll be willing to take the convenient bit of food, or the artificial fly, that arrives to them on the drift, but they're less willing to move far for it. If you're fishing the top, chances are far less that they'll move all the way up there for a single bit of food, unless the run is shallow. If you're fishing the bottom, you have to be sure your fly is right down there in the strike zone. No matter what you use, dry, nymph, streamer, or wet fly, you need to cover the water carefully with it so that it comes close to the trout on at least one of its passes.

When trout are truly dour in a run, unwilling to move because nothing in their world is moving, tempting them into feeding, you almost have to hit them on the nose with your fly. Most of the time, you just have to get the fly, usually a deep nymph, within a foot or two of them. Because runs hold so many trout, once you've found the right kind of water, you should have no trouble getting your fly within striking range of at least a few of them.

If a particular species of insect is active, emerging, or making its way to the surface to emerge, trout in runs tend to key on them. The surface is not choppy like that of a riffle, so trout get a better look at your fly. They are more likely to reject a poor imitation and are almost certain to reject a fly presented in an unrealistic manner.

When considering fishing tactics for runs, the same three levels apply as in riffles: the bottom, mid-depths, and top. The three levels tend to be more distinct in runs. The bottom is farther from the top, and the mid-depth area is a more distinct region between the two. Trout feeding in a riffle can cover all three levels while they hold on the bottom. They are usually willing to move to anything that drifts near them or over them at any of the levels. In a run, however, trout feeding at one level might refuse to respond to a fly fished at another.

Tackle

Again, I would like to mourn the lack of a caddy. Without one to carry a variety of outfits for fishing different water types and depths within a run, or as you move from run to run or from riffle to run to pool, it's advantageous to carry a rod that fishes well in a wide variety of water types, rather than one that fishes well in only one narrow set of circumstances.

The best rod for fishing riffles on all but small streams, examined in the previous chapter, seems to be an $8\frac{1}{2}$- to 9-footer balanced to cast a 4- or 5-weight line. In runs, it's likely you might want to toss some larger and heavier flies when trout are not feeding visibly. If you're willing to trot back to the rig whenever you bump into a run you'd like to fish, you'll probably return to the water with the same length rod balanced to cast a 6- or 7-weight line. But if you're like me, you don't like to trot in waders.

If you're walking and wading, the best thing to do when you leave the rig is to select a rod that is properly suited to the size of the creek, stream, or river you're about to fish, not to any particular water types within it. You'll want to fish all the water types you come to as you move along the stream in the course of a day. The right outfit will let you do that.

I've already divided the world into creeks, streams, and rivers. I typically use a 7-foot, 4-weight rod on creeks, an $8\frac{1}{2}$- to 9-foot, 4- or 5-weight on streams, and the same rod on rivers, though I add an $8\frac{1}{2}$- to 9-foot, 5- or 6-weight streamer rod when I'm boating rivers. The boat can serve as that caddy. On a day float, I'll have a light presentation rod rigged to fish hatches with dry flies and emergers, a long and somewhat slow rod rigged for nymphing, and in some cases another faster and heavier rod rigged with a streamer to bang the banks between anchoring points.

Personality has more to do with rod selection than any of us would like to admit. Your rod choice will work better for you than mine will if it suits your personality better than mine does. If you're a one-rod person, buy one that suits the type of fishing and the type of water you like to fish most. Make it do for everything else, but arm yourself perfectly for what you enjoy doing most.

Most fishing in runs, at least most of mine, can be accomplished with a floating line. I like the control it gives. Tending the drift of the fly can be more important in a run than it is in a riffle. I still use more double-taper floating lines than I do weight-forwards, but I do use a mix of both. When I buy a new rod, I always try it with a variety of lines, and it usually tells me which it likes best. Often that's a double-taper; just as often it's a weight-forward. I let the rod choose the line. I do like the weight-forward to have a long, heavy casting section before it tapers down to light running line. The longer belly gives me more control for mending and tending the float of a dry fly, swing of a wet or streamer, or drift of a nymph.

If you'd like to add some minor weight to your vest for fishing runs, it should take the form of a spare reel spool holding a sinking-tip line, with a fairly fast, but not superfast, sink rate. This will get your nymphs down into the deepest parts of runs. Without one, there will be a few places you might not be able to fish effectively, especially with streamers. In truth, I rarely carry a sinking-tip line on moving water unless I'm in a boat and have that spare rod rigged. If I need to get deep, I do it with nymphs, adding split shot or putty weight on the leader and a strike indicator up by the floating-line tip.

The leader you use in runs should start out about the length of the rod. Fish it the way it is or add tapering sections and a couple feet of tippet to suit the size flies you'll be casting. If you get into a hatch and find trout selective to something tiny, add fine tippet; this will also make the leader longer, which is the perfect way to respond to the increased fussiness of the trout without getting the situation confused by getting into mystical leader formulae.

If you switch to the sinking-tip line, shorten your leader to four to six feet. Trout are not leader-shy when they feed on the bottom. It doesn't make sense to use a sinking-tip line to get your fly to the bottom only to have a long, fine leader buoy the fly back up in the water column, riding high above your line tip.

Fishing the Bottom

Trout in runs, holding as they do along the bottom when they're not feeding, are less likely to move all the way up to the surface to take dry flies. That is why nymphs are more effective as searching patterns, especially in runs more than three or four feet deep. The bottom is the most promising

Depth Charging

Sometimes the water is simply not readable because of lack of clarity from runoff or rain, and you don't have much chance of hooking a trout unless you're able to get to the bottom with what I call a "depth charge" line. In this case, I was on the Yelcho River in Chile and didn't have the means to go to another river or time to wait for the water to drop and clear. So I donned raingear against the steady downpour, armed myself with an 8-weight rod and Teeny T-300 line, with the leader clipped back to about five feet, and fished the big runs where such things as joining currents predicted that trout would be lurking down near the bottom, beneath the heavy flows. Using a large streamer with some nonlead wire wrapped around its fuselage and a tungsten bead for a head, I gave the sinking line and heavy fly plenty of time to sink before drawing the line tight and beginning each swing and retrieve. I didn't catch

many trout, but I was out there fishing and did bring a few to my hand.

It was a writer's junket, and I was one of the few with enough experience in Chile to know you never go there without a depth charge line. As a beneficial consequence, I was also one of the few who were able to catch any trout when the water was out. Every time I brought one to hand, I was quickly surrounded by photographers, also on the junket but with few trout to include in their photos.

When you fish a relatively unfeatured run, and find an obvious interruption to the current embedded in it, be sure to fish it thoroughly. In this case Skip Morris is casting a streamer across the broken water downstream from a boulder and swimming it into and out of the slower water that is most likely to hold trout.

level to start fishing in a run, if you watch the water a while and do not discover any rising trout, because that is where trout spend all of their time when they have not moved off their stations to feed.

If you hoist some rocks or do a bit of kicking around with a net and find one insect abundant, your nymph selection should be related to any predominant natural. If you don't see evidence that trout are focused on one source of food, the same generic nymphs that worked in riffles are just as effective in runs. It's no secret that a small nymph box filled with a few favorites will take trout in a wide variety of situations. As in riffles, if you rig with two nymphs, one of them should be size 10 to 14, large to medium, and the other should be size 16 or smaller. And also as in riffles, you'll find that you catch most of your trout on the smaller of the two nymphs.

Once you've chosen a nymph or two, imitative or generic, you need to choose between floating and sinking-tip lines. My simple rule for all fishing is this: Fish a floating line until conditions demand something else. Don't use the sinking-tip unless you can't reach the bottom with the floating line and a reasonable amount of weight on the leader. But switch without hesitation if your fly isn't fishing down where you want it. For me, in runs of all but the greatest depths or current speeds, that is rare.

I have encountered situations in recent years in which I desired to get a nymph, or less often a streamer, to the bottom quickly in runs six to ten or more feet deep, with heavy currents. I don't seek out these sorts of situations on this continent; though they do exist, I tend to look for other types of fishing. But such situations come to me, and I find myself in need of reaching the bottom in heavy water, most often on big rivers in Argentina or Chile, where my mobility is restricted by the fact that I don't have my own rig and must depend on somebody else to move me. When I do get into such a situation, I arm myself with a rod I use only for steelhead here: a 9½-foot, 8-weight with an almost brutally fast action. It's a casting tool. I load it with what I call a "depth charge" line, usually a Teeny 300, the heaviest and fastest-sinking line the rod will handle. It's an aside in this book on reading moving water, but I would never be without a depth charge line for my 6- or 7-weight rod when fishing big lakes. That same line does fine in average stream and small river situations when you need to get to the bottom quickly in a deep run.

To fish the bottom in most runs, the best technique is the upstream dead-drift nymph rigged with split shot and strike indicator. Use a weighted nymph and add weight to the leader when needed, which will be most of the time. Start with the indicator twice the depth of the water above the top nymph, but be prepared to adjust it up or down if the water gets deeper or shallower, faster or slower. Cast short, twenty to thirty feet, and let your nymph sink to the bottom as it drifts downstream toward you. Raise the rod and draw in slack line as the nymph approaches; lower the rod and feed out slack as the nymph departs downstream. When it has reached the end of its drift, lift it up and lob it back upstream, a foot or two out from the line of the last drift. Set the hook at any hesitation of the indicator. When you've fished out the part of the run you can reach with easy and controlled casts from the first position, wade upstream ten to twenty feet and repeat the process.

This nymphing method is deadly when you fish the prime holding lies of a run. You can place the nymph right on top of the trout. By casting short, you retain excellent control so that you can lead and coax the nymph into the best water. This is especially important when working around boulders, the most common type of prime lies in a run.

Another method for fishing the bottom of a run works well where the water is fairly slow and not more than about four feet deep. It doesn't work if the water is either deep or swift. It calls for a sinking-tip line and a large nymph or streamer. Otherwise it's similar to the wet-fly tactics described in the previous chapter, though there can be a substantial difference in the size of the trout you catch.

Some big runs, such as this one on Washington's Yakima River fished by Skip Morris, have few features to tell you where trout might hold along the bottom. Your best choice in such a case is to read the water as well as you can and then fish nymphs in such a way that they cover all of the most likely water, giving you every chance to find trout in hidden lies down there.

The first part of this sinking-tip strategy calls for wading into position, in reference to a prime lie, so that you can cast slightly upstream from straight across and let the nymph sink and drift downstream toward the lie. Quickly toss mends into the drift so the line tip and fly sink without hindrance. The first twenty feet or so of the drift should be dead. By the time the fly is at a forty-five-degree angle downstream from you, it should be near the bottom and sweeping over or even into the prime lie. Stop following the drift with your rod at this point; let the line draw tight and begin to tug the nymph or streamer across the stream. The idea is to keep the fly down as deep as you can while drawing it slowly around until it is straight below you.

I used this strategy once while fishing the Williamson River behind the late Polly Rosborough, author of *Tying and Fishing the Fuzzy Nymphs*. Polly was celebrating his eighty-first birthday. He fished a few hours, finished up in a short run, and sat on a log to take a deserved rest. I caught up with him and asked for the honor of taking a few casts with his rod. He allowed it.

The stout Fred Thomas bamboo rod was armed with a sinking-tip line, short leader, and large but unweighted nymph of Polly's own devising, an experimental tie that he never did put into production and therefore did not attach a name to it. The run was about fifty feet wide, four feet deep, and no more than seventy-five feet long. Its length was lined with invisible ledges and trenches eroded into the bedrock bottom. I waded ten feet out, near the head of the run, to work downstream. I kept my casts to about fifty feet, placing each on the water a bit upstream from my position in the run. I tossed mends into each cast until the fly was as deep as I could get it. Then I fished it around on a slow, deep swing. Between casts, I took two or three steps, letting the current push me downstream. I didn't fish any particular lie because all the water looked good.

I'd taken only about ten casts and was still getting used to the feel of Polly's rod, when I felt a resounding thud. The trout came up and tumbled once in the air, then fought sullenly along the bottom. I did not encourage it to come up again; I had seen it and was afraid I might lose it. When it finally tired out and came to my hands, I estimated its weight somewhere between five and six pounds. I released the trout and handed Polly's rod back to him. It was a privilege to catch that trout on the famous man's rod.

There is another way to cover the deep ground in runs. It's usually used on the big rivers of the West: the Yellowstone below the park, the Missouri between dams, and other rivers of slightly smaller size. This method calls for casting with thirty-foot shooting tapers of super-fast-sinking line, the depth charge I mentioned earlier, backed by a hundred feet of Cobra or Amnesia monofilament, flung by 9- and 10-foot rods in the 9- and 10-weight class. The ammunition is large streamers: sizes 2 through 4 Woolly Buggers, Bunny Leeches, or Muddlers, some tied with tungsten beads, all weighted until they land with a thump when dropped into your palm.

The method is simple: wade deep, cast long and slightly upstream, allow some time for the streamer to plummet, then let the line tighten against the streamer and lead it around. Sometimes the retrieve is activated with a pulsed rod tip; sometimes the streamer is allowed to swing around with only the action given to it by the current. This method can be brutal to those who are not used to casting long with big gear. But it subdues many of the largest trout that are taken from such water each season.

Fishing the Mid-depths

Fishing the mid-depths in runs should be a tactic reserved for those few times when insect activity dictates it. When caddis pupae or mayfly nymphs are on the move from the bottom to the top, trout in runs sometimes feed on

The line of shallower and somewhat choppy water that feeds into a large run might be considered a riffle or part of the run itself. However you name such transitional water, fish it carefully, because it's far more likely to hold trout than the more brutal water downstream.

them in between. More often, these insects are taken by trout just as they leave the bottom or when they reach the barrier of the surface film. Except in rare circumstances, the mid-depths are barren water in a run. But it helps to know how to fish this level when trout are holding and feeding there, and you can see them doing it.

Wet-fly tactics for runs are largely those described for fishing the same types of flies in riffles. Use a floating line and leader the length of the rod or a bit longer. Cast wet flies, soft-hackled wets, or even small nymphs from an upstream position, letting them swing slowly around on the current, mending line to slow the fly and achieve some depth. Or fish slightly weighted or beadhead nymphs with upstream casts, using a strike indicator placed about the depth you want to fish your fly and small split shot or putty weight, if necessary, to get the nymph just to the depth of trout holding at midlevel.

I don't want to dwell on methods for fishing runs at mid-depths. Some sort of insect activity should lead you to this kind of fishing. You almost have to be able to see the trout suspended between the bottom and top, or at least suspect their presence there based on what you know of insect activity, before you should rig to fish this way. The tactic you select should then be based on what the insects and the trout are doing. Whether you use a wet fly or a nymph, present the fly in a manner that makes it arrive in front of the trout the same way the active natural insect would present itself just before it gets consumed.

Fishing the Surface

The top of a run is also a zone where you will do your best when you can see trout actively rising. The surface is smoother in runs than it is in riffles, and trout tend to be at least slightly more selective. Your choice of fly pattern and presentation should be based on what the hatch of the moment requires. Rather than going into detail and doubling the length of this already long chapter, I will refer you to my earlier book *Handbook of Hatches*, which covers in detail recognition, fly pattern selection, and presentation methods for all the aquatic insect hatches.

In runs from two to four feet deep, trout can often be brought to the surface with searching dry flies, as opposed to imitations, when there is little or no insect activity. You can enjoy some success fishing the dry fly without waiting for signs of rising fish. I call this *searching* fishing and enjoy it a great deal. I enjoy it more, in fact, than I do nymph fishing down on the bottom. But the latter method is likely to be much more productive than the former in the average run that lacks active, rising trout. Catching trout on nymphs is always more fun than not catching trout on dry flies.

Certain kinds of runs predict the success of the searching dry fly. They are generally shallow and broken, with visible or invisible boulders that form scattered holding lies in which trout feed opportunistically. If the stream is narrow and forested, the chance of terrestrial activity on the water is increased, and trout are more likely to be alert for opportunities that arrive on the surface. In late summer, hoppers and other terrestrials cause trout to look upward in meadow streams.

Fishing runs with dry flies, when trout are not rising, can be done with the same dressings used on riffles: Elk Hair Caddis, Royal Wulff, Humpy, and various traditional dressings such as the Adams and Light Cahill in size 12 to 16. Meaty attractor drys seem to work best when you have to draw trout up through several feet of water.

Presentation techniques for dry flies in runs are not any news. In most cases where the water is at least slightly wrinkled on top, you can fish upstream casts without fear of frightening the trout, though it's always best to take a position off to the side at an angle, so your cast does not line the trout you'd like to catch. Curb the distance of your casts so you can control the drift of the dry and keep it from dragging on the surface. The farther off you fish, the more chance the line and leader will cross unseen conflicting currents, causing the fly to move unnaturally, but so subtly that you don't even notice it. Trout always notice such microdrag, however, and are always put off by it.

If the run is particularly smooth on top, you might find it advantageous to fish reach casts that are almost straight out from you, the line cutting at an angle across the currents. Mend and tend the line so your fly has a long, free drift. This keeps the line and leader from crossing the trout before the fly reaches it. It's an exceptionally good tactic for stalking pinpointed prime lies, presenting your casts to the most likely holding water in the run. It's also very effective for the most readable trout of all: those exposing themselves by rising to take insects off the surface.

Anatomy of a Pool

Show almost any fisherman a likely lie and he will declare, "That's a nice-looking pool!" It might be a riffle or run, but if it looks like water that holds trout, it's automatically called a pool. Perhaps this terminology serves a purpose by defining likely holding water and is not mistaken. But to make a book about reading trout water work, a pool must have its own definition, separate from riffles and runs and other water types that hold trout.

A *true pool* is a reach in a creek, stream, or river where the gradient flattens, the water slows nearly to stillness, and the bottom deepens compared with the water upstream and down from it.

True pools—bodies of deep, slow water, not just any water that looks as if it might hold a trout—are often passed up by fly fishermen, especially on big rivers, because they are the most difficult water to fish. But they offer maximum shelter from currents and protection from predators. Pools often hold the biggest trout in any stream. They are well worth the effort it takes to get to know how to read and fish them.

STRUCTURE

A pool varies in size depending on the size of the stream that forms it. A pool in a mountain headwater creek might be ten feet long, five feet wide, and just three or four feet deep. That's not large, but it's certainly large enough to give rest and comfort to trout of the size found in a creek that hurries off a hillside. Some pools in the Big Hole River in Montana, on the other hand, are so wide they cannot be cast across. They might be a hundred yards long, half that wide, and ten to twelve feet deep. Such pools hold trout that suit their sizes.

A pool, small or large, is the slowest and deepest water in the stream in which it's found. Nearby water types, upstream and down, have steeper

Classic trout stream pools have a distinct anatomy, with the water entering at the *head* in the form of a riffle or run, then slowing and deepening in the *body*, and finally lifting and speeding up again in the *tailout*, which generally leads into the next riffle or run downstream. Pools vary from this design far more often than they adhere to it, but they're all recognized by being deeper and slower than the water upstream and down from them.

This pool, in the Black Canyon reach of the Yellowstone River, follows the prescription more precisely than most. It has a set of rapids that loses speed and turns into a run at its head. The body spreads and slows and becomes quite deep, but it keeps enough current that its water is far from still. The tailout in this case moves through a boulder garden run before gathering speed and tipping toward what is probably the next set of rapids downstream, right around the corner.

gradients and are always faster and shallower. A pool need not be a sudden dramatic drop in the stream's bottom. There can be a gradual deepening, then a long stretch at an even depth, followed by a gradual rise once again. But if the bottom depth were a graph line drawn through the course of a stream's water types—riffles, runs, and flats—a pool would be a distinct dip below and rise back up to that line.

The structure of a typical pool is simple: It has a *head* where the water enters, a *body* that deepens and darkens, and a *tailout* where the water lifts and shallows out again to break over into the next water type. There are as many variations as there are typical pools, but most pools have these same three parts in one shape or another.

Some pools are long and deep. They might be classified as runs if the water were not so slow, almost lacking a defining current. The demarcation line between a slow run and an even slower pool is far from distinct. At times one grades into the other, and it's difficult to tell where one ends and the other begins. In fishing terms, it matters little, since you'd fish that bit of water with the same tackle and tactics no matter how you defined it.

Other pools are short and abrupt, dropping straight off into deep water. Because the bottom drops away, the current loses most or even all of its thrust. Sometimes it eddies back on the sides, or even upward and backward, before shallowing up and gathering speed for its trip downstream once again.

True trout stream pools are most common in creeks and streams that are still up in the mountains and foothills, eroding their beds into immature streamcourses, trying to subdue the land. Some of the substrate over which such waters flow is more resistant than that farther down the valley, out into the flatlands. Flowing water in the headwaters has had fewer eons in which to grind big boulders into smaller rocks, pebbles, and sand. Bedrock is more common, and the typical bounding creek or stream digs lots of holes into it. There isn't as much fine material for the current to shove around, fill in the pools, and even things out—to draw that graph line level.

A typical mature river, lower down in the system, flowing across land with less gradient, has lots of long, slow runs but fewer true deep pools. A pool in a large river will dig itself deeply into that softer bottom, however, offering the largest territories and most comfortable stations to the biggest trout in any system. As the foods available in such placid waters can take the form of chubs and suckers and smaller trout rather than smaller insects, you'll find occasional predaceous lunkers in true pools in the most mature streams and rivers. If you're a trophy hunter, or just happen to be the sort of person who enjoys catching large trout rather than small ones, you should learn to read pools and pay careful attention to them when your travels after trout take you to them. Don't pass them by.

The head of a pool usually has some rubble and rock tumbled into it from the riffle or run upstream. This makes at least fairly good habitat for aquatic insects, and these insects are also delivered into the head of the pool from whatever water lies just upstream. The bottom beneath the head of a pool is similar in structure to the bottom of a run, and the trout foods that live there tend to be larger types similar to those found in a run.

The bottom structure of the body of a pool depends largely on the kind of stream it is in. Pools toward the headwaters are often scoured to bedrock. The finest sediment—pebbles, sand, and even silt—is always deposited wherever the water is slowest, so pools lower down in stream systems, where they aren't worked to bedrock, tend to have bottoms that lack cobble and therefore the myriad small spaces between stones that would allow aquatic insects to thrive in great numbers. Instead, pool bottoms are often littered with larger rocks and boulders embedded into finer substrates. As you would suspect, food in the depths of big pools tends away from aquatic insects and toward larger beasts such as crayfish and baitfish.

The tailout of a pool is similar to that of a run. It's a buildup of current over a bottom of small pebbles and stones, with some larger cobble. Unless the stream or river lacks vigor, the tailout gathers speed as it lifts up, and this increased flow keeps the bottom gravel cleansed of silt. The stones can be as clean as those found in a riffle. Sunlight strikes down to them through the shallow water, increasing photosynthetic growth. Pool tailouts are excellent habitat for many of the same kinds of smaller aquatic insects that live in riffles. Trout of normal size move onto tailouts to feed on the insects, especially during hatches. Trout of predatory size move out of the depths and onto tailouts to feed on the things that feed on those insects.

NEEDS OF TROUT

Pools can meet all the needs of trout in ways that make them satisfying to the very largest trout found in a given creek, stream, or river. They can be the primest of prime lies. If you're after the largest trout of your life, pools should rise to the top of your list of places to fish. But you have to assess a pool's nature carefully to see if it provides everything a trout might need.

Many pools offer maximum shelter from currents and protection from predators but have little food to offer their trout. Such pools become holding lies for the largest trout in a stream, from which the trout venture out onto feeding lies, or bomb shelters where trout can seek shelter when they're frightened off nearby feeding or holding lies.

If a long, deep pool, especially one dug out of a bedrock bottom, is barren of food, it will often be barren of trout unless they're driven into it. If you're able to assess a pool as one that provides good cover but little feed, you'll be able to confine your fishing to the most productive parts— the head and the tailout—and relegate the rest of it to the category of empty water, saving the time you might spend fishing it.

Shelter from Currents

Pools offer trout the maximum in terms of shelter from abusive currents. The water is slow. In its depths, trout find places where they can expend little energy to hold a station, at least when water flows are normal.

Shelter from currents becomes a live-or-die situation during winter spate and spring runoff. Trout find shelter as near their home lies as they can, but sometimes little is available. When conditions are extreme, many trout are pushed off their territories and into pools. Perhaps that is why nature, with her patient wisdom, turns down the territorial instinct as water gets slower. Trout can leave their scattered territories in riffles and runs and gather in shelter in pools, with fewer quarrels and less stress in an already stressful situation.

Protection from Predators

The need for protection from predators is met in pools by depth and darkness. Trout are secure there from aerial predation. They are not entirely secure from the likes of mergansers and otters and us. But safety is relative: Trout are safer in a deep pool than they are in a run where the light is better, as is the chance for a surprise attack.

The darkness at the bottom of a pool often makes it the best bomb shelter for hundreds of feet on a small stream, hundreds of yards on a big river. It's where trout flee when danger looms. Trout do not always come double-timing in battalions from distant lies. A trout on a long flat will not often flee the length of it to bury its head in a pool. It will have a smaller hide nearer home, perhaps doing no more than burying itself beneath rooted plant growth that obstructs its outline and makes it more difficult to spot from above. But a pool adjacent to a flat or riffle might be sanctuary for most of the trout that have found holding or feeding lies in the nearby shallow water.

That is why you often find trout in less-than-comfortable depths in association with nearby deep water, but you often fail to find trout in the same sort of water that is distant from any sort of pool they can use as a bomb shelter.

Trout Foods

Trout in small to medium-size pools are fed by a combination of items delivered into the pool from whatever sort of water is upstream from it, usually aquatic insects, and things that drop to the pool from streamside vegetation or overhead forest canopies, usually terrestrial insects. On creeks and small streams, this can cause trout to shift their stations from the upstream ends of pools early in the season, when aquatics are busy hatching and being delivered on the entering currents, to the lower ends of pools late in the season, when hatches are mostly over but terrestrials begin making unhappy visits to streams, usually landing lower down on the body of the pool.

The need for food in big pools is most often met in the form of the biggest bites. Riffles offer a constant supply of small to medium-size aquatic insects to their trout. Runs furnish a slightly sparser supply of small to slightly larger insects and some crustaceans such as crayfish. Big pools provide some aquatic insects, usually on the heads and tailouts, and some terrestrial insects, usually just along the edges. The fare turns more to baitfish in the body of a big pool; crayfish and sculpins scuttling in the nearby shallows draw trout out to hunt as light falls at dusk, through the night, and again at dawn. On cloudy days, and even during rainstorms, you can sometimes find big trout out taking the sorts of risks in shallow water that they would normally take only under cover of night.

The increasing size of the prey is probably a part of the reason that trout move through a simple progression as they grow older and larger, establishing successive territories first in riffles, then in runs, and finally in pools. But the largest part of the progression involves the relationship between the value of a territory and the amount of shelter it gives from pushy currents. Pools offer the most of that. No matter the size of the stream, you'll find its largest trout in its deepest territories, or at least in near association with them.

When any trout takes up residence on one of the best territories in a big pool, it's usually big enough to take advantage of the larger food types offered there. It's also usually large enough to swat smaller trout out of its way.

The bottom of a pool is generally poor soil for aquatic insect larvae and nymphs. But trout in a pool are the recipients of hatches that occur in the riffle or run upstream from the pool. The drift at the head of a pool can be extremely rich and can cause large trout to nose up out of the depths to feed, even selectively. Hatches are also excellent at times on tailouts. If a hatch is heavy on a tailout, even if the insects are tiny, the largest trout in a pool will sometimes back down with the current to hold lazily and pick insects drifting above the shallows. At times, big trout will risk predation in the shallow and clear water of a tailout in order to feed.

When selecting flies to match the foods of pools, in the absence of a hatch or some other obvious sign that you should use an imitation, it's best to match the biggest bites the trout might find in that bit of water if you desire to take its largest trout. You want to use something with sufficient size to stimulate the urge to feed in a trout that might have just eaten and will not move far for the next meal unless it's a big one.

Big trout are not always active in pools. They tend instead to chase down a large meal, then spend a few hours to a few days digesting it. The larger the trout, the less often it feeds, assuming it gets the kind of large meal it likes. That's why you often have to get your enticement down deep and moving slowly before a big trout will take it.

Temperature and Oxygen

If there's a bad place for a trout to be in any size stream when temperatures get high and oxygen levels get low, it's in a pool. The water is still, or nearly so. The surface of a pool is not broken, as it is in rapids or a riffle. There is little exchange with the air to freshen the water. That doesn't necessarily mean trout will move out of a pool when temperatures go up. In a pool several feet deep, depth in itself is some shield from heat.

If conditions approach distressing, however, trout metabolism will slow and nearly cease to function. A pool is a good place to do that, because it has little current to fight, and trout can put things on hold without getting pushed off their stations and out of their territories.

When conditions become severe, trout will sometimes move out of pools and seek riffles and even rapids, hanging in pockets surrounded by white and therefore well-oxygenated water, where they would not normally be found. When temperatures are high and oxygen is low, there are better places to fish than pools. Certainly, if the trout are stressed, you should not fish for them there or anywhere else.

Exceptions to all the above occur in pools that benefit from underwater springs. These are not rare, especially in freestone streams, but they are also not easy to find. If you have a home stream, one that you fish over and over, and it's susceptible to warming, it would be worth your while to spend time dipping a thermometer into the water as you move along, whenever you fish, looking for obvious diversions from the average. If you find a cold spot, you will have found a trout hot spot.

A cool side-stream freshet creates the same conditions, cooling the main body of the stream it enters and making the nearest pool downstream a refreshing place to hold. It's good to notice the arrival of feeder creeks along the course of any stream as you fish. When conditions begin to trend toward bad on the main stream in the heat of summer, you can try the pools immediately downstream from the feeder. If the water of the feeder is cool enough and its volume sufficient, its freshening effect might continue for several hundred feet. The smaller stream might also draw trout to move out of pools in the main stream, and you'll be most likely to find those trout, especially the outsize ones, in pools of the smaller feeder.

TYPES OF POOLS

Moving water has many different kinds of pools, defined largely by the way currents create them. One type of pool grades into another. Some pools are combinations of more than one type. Different stream types tend to have specific pool types; for example, mountain streams tend to have plunge pools, foothill streams to have classic pools, and big rivers to have bend pools. But any size creek, stream, or river can have any type of pool.

Classic pools are places where the gradient of the watercourse levels off and the current, usually at high water in winter and spring, has had time to dig the bottom here deeper than upstream or downstream. The flow simply pauses to rest. Such pools are wider, deeper, and darker than the water all around them. The pool can have some significant current or be nearly as slow as a farm pond. Classic pools have the simple anatomy already described: head, body, and tailout.

Large classic pools are relatively rare on most rivers, though it might seem they are common until we close our eyes and ransack our minds to think of where we've seen them. There are none on such famous rivers as the Deschutes and the upper Madison. But they are important features on the Big Hole, many famous Catskill and Adirondack rivers, and other streams all up and down the Appalachian Mountain chain, where water has had plenty of time to erode the traditional trout stream structure of riffle to run to classic pool, then back to riffle and run and pool again.

This classic pool, on California's Sacramento River near Dunsmuir, enters in a cascade, broadens and deepens in a long body, and, though it's not shown in the photo, becomes shallower and gathers speed to plunge into the next cascade downstream. Its main current and therefore its deepest depths are pushed against the steep hillside on the far side. Trout would be found the length of the pool there, under those currents, because that forms the sanctuary water. This pool fits Charlie Brooks's description of a "bomb shelter."

Classic pools are much more common on smaller foothill streams than they are in the headwaters or out in the flatlands of mature rivers. Chances are your home water has them, and chances are you fish their tailouts sometimes, their bodies almost never, and their heads almost always. If you're after big trout, that order might benefit from being shuffled, since the largest trout in a pool holds on its tailout at rare hatch times, in its body almost always, and at its head only when feed is being delivered in abundance from upstream.

Bend pools are found where the river makes a change in course. The force of the main current sweeps around the outside of the curve, eroding it away, digging it deeper. The flow of a bend pool is slow, but it seldom loses its definition as the main current. In some ways, bend pools border on being runs, and many runs force their currents to one side, always the outside. But bend pools are distinguished from runs by the depths they achieve—three to five feet on small streams, four to ten or more feet on larger streams and rivers—with shallower water leading in at the head and out at the tailout.

This small but typical bend pool, fished by the late Col. Tony Robnett on Soda Butte Creek in Yellowstone Park, pushes its current against the bank, eroding its greatest depths there. As in all but the deepest and slowest pools, trout are found under that main current, wherever they can find a break from it, because the current delivers the most food, and the trout are also protected from predators here.

Most bend pools are broad and sweeping, with shallow gravel bars on the inside of the turns and deep slots cut on the outside, pushed up against the far bank. The deeper bank is often defined by boulders or even riprap, because it takes such structure to prevent the bank from simply sloughing away. Bend pools with soft outside banks often have similar characteristics to those with firmer banks, but they shelter fewer trout unless undercuts and indentations have eroded into them, providing the fish places to avoid the current. The boulders of a structured bank offer trout prime lies, where they have breaks from the current, enough depth to provide overhead protection, and the constant current to deliver food.

Almost all trout waters have some bend pools, but they are most common in medium to large freestone trout streams and rivers. Meadow streams, with their meandering courses cut through level river bottoms, also have cut-bank bend pools sprinkled at almost regular intervals, wherever they take a turn.

Cliff pools are bend pools where the water arrives at a fairly abrupt angle against an immovable object: a giant boulder, cliff, or solid high bank

This perfect cliff pool, in Montana's Dearborn River, descends as a riffle, then is slowed and turned by the sharp rock face. It promises to be more productive of trout than most cliff pools, because that rich riffle would deliver far more food than the average set of currents entering a cliff pool.

that the water has worked an indentation into. The flow stacks up before making its turn, and the main current then slides along the cliff. The result is a deep undercut eroded at the turn, forming a short, deep pool and then a runout with the deepest water eroded right against the cliff or high bank. Trout hang out deep under that entering current where it's pushed against the cliff. Other trout line out under the main current where it trails along the cliff.

You'll find the largest trout in the deepest water toward the head of a cliff pool, where the food enters and shelter from currents and protection from predators are also greatest. That's the prime lie, and it will usually hold the biggest trout. As you move downstream, lies provide less food, slightly marginal shelter from currents, and less protection from predators. In a cliff pool, expect trout to become smaller toward the tailout, larger as you move toward the deeper prime lie at the turn.

Because cliffs are a function of younger geography, you'll find them most often in mountain creeks, less often in foothill streams, and very rarely in mature low-gradient rivers.

Ledge-rock pools form wherever the current has worked its way to solid bedrock and then eroded a deep channel down some length of it.

This ledge-rock pool is in a classic trout stream, Cedar Run in Pennsylvania. The water enters as a current tongue from a brisk riffle; deepens out along the ledge on the far side; carries its main depth, as almost all pools do, directly under its main current; and, though it's not shown in this photo, lifts toward a tailout and another riffle downstream. It's easily read. Trout would be found under that current tongue and in darkness of the depths against the ledge. Cedar Run is famous for holding a few nocturnal brown trout. This would be the perfect place to find one, though during daylight when we fished it, the big one would not be expected to come out and attach itself to a fly.

These pools tend to be abruptly deep, very narrow, and banked in tightly on both sides. They are often features of a fairly steep geography and are found toward the upper end of a stream system, where it still rushes through the mountains and high-gradient foothills. But ledge-rock pools can be found along any stream where the bottom is bedrock. Such pools are common in low-gradient streams where the bottom consists of basalt flows or the stream has had time to work its way to ancient underlying granite or sedimentary stone.

Ledge-rock pools are abundant on the low-gradient Williamson River, where Polly Rosborough did his research for *Tying and Fishing the Fuzzy Nymphs*. He fished the annual migration of large trout that moved upstream out of Klamath Lake, with their life histories of summer steelhead cut off by the lake from the traditional route to the Pacific Ocean. Polly cast his big nymphs and streamers long, on sinking-tip lines, over the many ledges and trenches worn by water into the basaltic bed of his home river, where those big trout paused to rest on their migrations to spawning beds in tributaries farther upstream.

The same types of ledge-rock lies are cut into the ancient bedrock of the Davidson River, where I fished it with Kevin Howell near his home in Pisgah Forest, North Carolina. Kevin is coauthor with his late father, Don Howell, of *Tying and Fishing Southern Appalachian Trout Flies*. I hoisted a few rocks off the bottom and found several fish fly larvae, so closely related to hellgrammites that they're usually mistaken for them with no deleterious results in terms of imitations, crouching in crevices eroded into the faces of the old, worn stones.

Trout held in similar, though much larger, ledge and trench lies eroded into the streambed. Kevin showed me how to tempt those trout out with size 12 Beadhead Bird's Nest nymphs fished under white yarn indicators, on long tippets of a single diameter, allowed to sink nearly to the bottom in those trenches. The trout had learned to be wary of indicators in bright colors, but the white was not much different from the bits of foam that they saw continually. Whenever our indicators moved in opposition to the tiny patches of foam all around them, it was easy to decide what to do, and the results usually weighed two or three pounds when led to hand.

Plunge pools are often strung like fat pearls down the course of a steep-gradient mountain stream. They form wherever water tumbles over an obstruction, one plunge and then pool after another. The water might make a one- or two-foot drop over a miniature waterfall, eroding a pocket downstream, or it might plunge down from several feet, forming a waterfall pool. The pool might be brief and frothed at the head, dark in the depths and just a few feet long, or it might be wide and spreading and deep.

In my experience, plunge pools that form holding lies slightly larger than those typical for the stream also produce trout slightly larger than those average for the stream. Larger plunge pools formed below tall waterfalls, on the other hand, always look better but produce less, at least for me. It might be related to a lack of abundant feed where the falls eroded the bottom down to barren bedrock. It's more likely related, however, to my propensity to fish dry flies on small streams and my reluctance to rig the type of bottom-bouncing nymph that could be cast right to where the waterfall plunges in, the downcurrent of which would drive a heavily weighted nymph to the bottom. The largest trout might be right there, lurking in safety. You'll have to be the one to rerig for that single waterfall pool and fish its depths. Tell me how you do.

Plunge pools also form below waterfalls in low-gradient trout streams and rivers, farther down the system. These are always perfect places to take

A plunge pool, as its name implies, is formed where a miniature waterfall dives in, gouges out some abrupt depth, gathers itself, and plunges over the next tiny falls to form another plunge pool. Plunge pools are typical of mountain streams, where the water has to step down in a hurry. They never form where the gradient is less than steep or the streambed is consistently soft, lacking boulders and base rock.

Higher waterfalls, six to sixty feet, also form plunge pools. These are, without variation, deep and dark and beautiful. Photographers love them. But I usually find them somewhat lacking in trout. It's possible that they're oversaturated with air or with nitrogen. It's also possible that when I encounter such a tall waterfall, I fail to switch from the light gear with which I'm typically fishing my way up a small stream and therefore fail to get my flies right to the bottom in the suddenly deeper water. That's where the trout would be, if they're down there.

pictures of folks fishing, but they are rarely any more productive than the riffles, runs, and pools up and down the rest of the stream or river, again for me. Rivers that have upstream spawning migrations, however, gather frustrated trout in waterfall pools. These can be excellent places to fish if you don't mind joining the occasional crowd and casting over thwarted trout.

Eddy pools are common on most rivers. These are places where a strong current swings out from the bank and then takes a slow whirl back upstream to fill the void its passage has left. The water along the side of the main current shifts into the space next to the bank. It finds a vacuum there and noses back upstream to see where it's been. When it goes full

An eddy can be slight to monstrous. I fished one once, on the Futaleufu River in Chile, that was bigger than some still-water ponds I fish. Large rainbow trout cruised its cliffed edges, willing to gulp Woolly Buggers dangled in their paths so long as you kept out of their sight while you dangled them. But it wouldn't be fair to use this as an example, because it's the only one of its size I've ever seen and the only one that had oversize trout working its back-currents. It's far from rare, however, to see the largest trout in any stream or river feeding in its eddies.

This eddy on the Deschutes River, fished by Robert Sheley, is much more typical in size and shape. The large rock jutting out to Robert's left shunts the current outward. It makes a full circle around him and comes back to join the main current almost at his feet. That patch of foam in front of him, from which he has already extracted a couple trout before moving into position close to it to fish downstream from it, is the prime lie in the eddy. Any insects either hatching or falling to the water would eventually be brought there by the circling currents. Trout know that, so they rest under the foam, rising to poke their noses into it whenever they see something they'd like to pick out of it.

The reversed currents along the shoreline, coming upstream toward Robert, form the second most prime lies. Trout hold there, waiting for the same currents to deliver food toward the foam patch. They're facing upcurrent, which in this case is downstream, so Robert is presenting his dry fly to them with an upstream cast, though he's looking downriver to do it. That's the way it should be done.

The seething currents in the center of the eddy, where the water upwells, would be almost impossible to fish. A dry fly would get instant drag. If trout are rising there, you won't have much chance against them unless you switch to a soft-hackled wet or small nymph.

circle and joins the main current to go downstream again, it has left an eddy in its wake.

The resulting backwater has a permanent circular current. It forms a gentle place for trout to hold, with depth and therefore excellent protection from predators. As a bonus to trout, an eddy is a trap for all the debris and drift that come down the current, which often include big numbers of small insects that either emerge in the eddy or accumulate there. Trout can hold anywhere in an eddy, from its reversed edge currents to its eye, feeding quietly on aquatic insects and terrestrials and anything else caught on the conveyor that goes round and round. Because eddy trout are picky and eddy currents are conflicting and confusing, they can be difficult to fish. It's important to remember that the current along the bank goes upstream, not down, and you might have to position yourself exactly the opposite of the way you'd think right in order to get the types of drag-free drifts necessary to fool trout in such water.

Because eddies are always formed in relation to stream- and riverbanks, methods for fishing them are covered in chapter 10.

HOLDING LIES

The primary holding lies in a pool are analogous to the parts of the pool: the head, body, and tailout. Trout are found in one of those places in a pool most of the time. But any obstruction that slows whatever current might move through a pool, and also offers a place to conceal a trout from overhead, will form an excellent lie.

Trout often hang along the edges of a pool, especially one that keeps its depth right to the bank. Terrestrial insects drop to the surface there. If trout can find any sort of safety, in the form of undercuts, overhanging branches, or just shade, they might tuck in and feed opportunistically, taking whatever lands on the surface of the pool.

Head

The head of a pool usually takes the form of a current tongue rushing or plunging in from a riffle or run upstream. It's a prime place for trout to hold. The sudden deepening of the pool slows the water, providing shelter from the current. Depth and the broken surface of the current tongue give protection from predators. The current delivers food in the form of aquatic insects drifting in from the faster and shallower water upstream.

One or two little corner pockets form at the head of any mountain or foothill pool, wherever the current makes its way between boulders and

The head of a pool, such as this one on Kelly Creek in Idaho, always enters in some form of a current tongue, delivering most of the food into a pool, and you need to fish it carefully. The shape and size of the pool, along with the conditions of the water and weather, will dictate how you should fish it. If trout are up and feeding actively, or even sporadically but on the surface, then fish it with dry flies. If they're not, and the weather is not comfortable for you, consider that trout are more likely on the bottom and don't have any reason to leave it, and you need to get a nymph or streamer down there to root them out.

enters the pool either dropping or rushing past solid rock banks. Some pools have such lies on both sides, some have them on just one side, and some, entering over gravel, have none. These corner pockets form only in pools with sufficient depth at the head to cover trout from overhead predation. They usually take the form of very small, almost still eddies over dark water. On the tiniest headwater stream, a corner pocket with a surface area the size of a basketball will hold a trout as large as any other lie in the pool might offer. Larger streams have bigger corner lies and usually hold at least one or two nice trout.

The most common trout lies at the head of a pool are generally down on the bottom, directly beneath the current, or tucked under each side of the current tongue, along the seam between fast water and slow. This water is very similar to that in riffle corners and holds trout for the same set of reasons: It's the best place for the fish to shelter from the current, take cover from predators, and still be in position to spear out to take advantage of all the food that a rich current trots by.

Often a pool is formed by a shelf or drop-off at the head, followed by water that is deeper but of uniform depth down to the tailout. Where this is true, the best holding water in the pool is located in that first few feet of interrupted water, right under the entering current, since the most shelter, protection, and feed are all located right there. The rest of the pool, with its even depth, will hold trout. But the prime lie will be at the head of the pool.

Trout will sometimes shelter in the main part of a pool, moving up to the head only when they want to feed. That will be true especially if a hatch happens; they'll always respond. It will also be true if the water at the head of the pool is too shallow and clear to provide continual protection from predators. Then trout will only respond to an opportunity to feed. The rest of the time, they'll hold in the deeper part of the pool, where they'll most likely be willing to accept a nymph fished at their level but unwilling to rise out of the depths of the pool and go all the way to the top for a dry fly.

Body

In a typical pool, the water depth slopes off either slowly or abruptly downstream from the head. The body of a pool thus formed might not always hold the greatest concentration of trout, but it's likely to hold the largest. They lurk in the quiet depths, waiting for a chance at something worth eating, letting trout that are still gaining their growth stay up at the head of the pool to argue over tidbits that arrive on the drift. When the time gets right, these big trout might nose up toward the current tongue and chase the small trout that usually feed there, or drop down toward the tailout and pursue the same sort of sport.

The body of a plunge pool in a small mountain creek usually holds one dominant trout. Others might rush your fly in peripheral waters; the largest trout will usually occupy the lie that offers it the first and best shot at whatever food the current tongue entering the pool brings. That is why it's often best to stalk these tiny pools carefully and present your first cast to the most likely holding water, rather than fish your way up through them. If you fish poor water first, you're likely to hook a small fish first. Its antics might spoil the pool before you can show your fly to the local bully.

The body of a medium-size pool might hold several trout nearly equal in size. Sometimes it holds a dominant trout, one that has established such a favorable territory that it has outgrown its competitors by what amounts to lots of heft when you get a chance to hold it in your hand.

It's easy to read the body of this pool, on the Big Thompson River, where Masako fishes under the watchful eye of guide Chuck Prather, out of Sylvan Dale Guest Ranch. The foam line on the far side, up against the trees, defines the precise line down which any drift would travel. The deepest water is there, and that's where any trout worth catching would hold, in order to take advantage of the shelter from currents, protection from predators, and delivery of food offered by the pool.

A pool in a large river might be the size of a farm pond, and it might hold many large trout. Each trout usually has its own territory and a particular station when resting. But the urge toward territoriality is relaxed as moving water slows, and it disappears when the water approaches stillness. These trout are hunters. They spend most of their time in torpor, or near it, digesting the last meal. When it's time to hunt, they forage throughout the pool, sticking to the protection of its depths in daytime, but moving up into the shallows along the edges, to the head or tailout, at dawn and dusk.

Tailout

The tailout delivers a pool to whatever type of water forms downstream from it. Some tailouts lift and then plunge into cascades so swiftly that they are not fishable. Most lift gently, thinning and fanning out, until they ease over into a run or riffle below. Tailouts have some aspects of riffles, with bottoms of fairly uniform gravel, usually rich in insect life. But the

Trout typically move onto the tailout of a pool only when food—in the form of hatching or falling insects during the day or prowling baitfish and crayfish at evening—attract them there. At other times, they remain in the deeper body of the pool, where they are safer and more comfortable.

In this case, on the Rio Emperador Guillermo in Chile, trout responded to a size 14 mahogany dun hatch, and we had fine fishing for an hour or two where the duns were swept onto the tailouts of pools. As with flats, you read tailouts by looking for working trout and normally bypass them if you don't see any trout feeding, unless you fish them in the low light of morning or evening, when you should probe them with streamers.

surface lacks the chop of a riffle. The result can be fast water that is also slick on top, a combination that leads to some tough fishing.

Trout usually do not hold on a brisk tailout unless some sort of insect or other activity entices them there. The water lacks sufficient shelter from currents and protection from predators to make shallow and swift tailouts anything but feeding lies. But they can send up aquatic insect hatches, and trout are often drawn to tailouts to feed.

Some slow tailouts have all of the aspects of rich flats. Their thinness promotes photosynthetic growth on bottom rocks, and they have prolific insect life. Their gentle flows do not require much more than a few stones or small boulders to provide shelter from currents. But the skinny water offers little protection from bird predation. I have encountered lots of trout holding in gentle tailouts, often right at the lip. These trout are rarely large, although there have been exceptions. Trout on tailouts are invariably wary, completing the analogy to flats, where they flee at the faintest sign of passing substance or shadow.

Shallow tailouts are excellent places for sculpins, baitfish, and crayfish to forage, because they're gathering points for leaves and other types of forest debris on which omnivores enjoy nibbling and where small predators find lots of immature insects to eat. Baitfish and crayfish usually become active in the morning or evening, when light is low. The same low light prompts large trout to leave the sanctuary of the body of the pool and come out to forage on the foragers. That's why you're liable to take small trout at the lower lip of a pool during the day but occasionally set the hook into something that has trouble keeping its dorsal fin submerged in the thin water of a tailout at dawn and dusk.

Obstructions

The head, body, and tailout of a pool are all obvious lies for trout at one time or another. But within these parts of a pool, tip-offs to specific holding lies are the same as they are in other parts of the stream. Any boulder or log or shelving area in the head of the pool interrupts the fast inflow of water and allows trout to hold in shelter from the current.

Large boulders in the head, body, or tailout of a pool are natural havens for all sorts of crayfish and baitfish and offer maximum shelter to trout at the same time they feed them. Logs and root wads are commonly caught in the slow water of pools, especially along the edges of bend pools. These offer special comfort to the largest trout. They are also among the toughest places to fish, which is one reason they hold the largest trout. In a stream with heavy fishing pressure, any hindrance to the angler is bound to be shelter for the trout, since fishing pressure is one of the major shaping forces in the activity of the trout. We become main predators.

Edges

When trout move out of the depths of a pool to go hunting in its shallows, sometimes they move toward the head or tailout. Other times they climb right up onto thin gravel bars, where they find baitfish and crayfish and even smaller versions of each other. Cannibalistic brown trout are especially prone to hunting shallow edges, but rarely when the light is bright.

Undercut banks, along the sides of any deep pool, are perfect sanctuary water. Many types of aquatic insects flit around in the darkness back there.

Terrestrial insects drop in from such edges and are easy to take without much exposure to the trout nosing out from its hiding place to make a kill. An undercut bank that is an extension of a deep pool will be a prime lie in that pool. Most of the time, however, such lies come under the cover of bank water, and these are treated in chapter 10.

Deeper edges of pools, where the water maintains its darkness and depth right to the edge, can attract trout in late summer and early fall, when terrestrial drop-ins become the dominant food, usually in small to medium-size streams. The banks might be grassy and provide a supply of beetles, ants, and grasshoppers. I've found lots of midsummer trout tight against forested small-stream banks, at the edges of pools, especially where they can tuck into the shade of an overhanging branch. They feed on beetles, ants, and inchworms that fall to the water. The advantage of the edge lie, as opposed to one in midpool, is obvious: A trout at the edge can race up and arrive first to food that drops to the pool's surface.

FISHING STRATEGIES

When thinking about tactics for pools, it's best to consider the anatomy of the pool and dissect it into its main parts—head, body, and tailout—rather than the three layers dealt with in riffles and runs: bottom, mid-depths, and top.

The head of a pool, where the water enters with a strong flow from the riffle or run upstream, is best fished with an extension of the tactics you were using when you arrived there, from the same riffle or run, or from a similar one if you approached from downstream. The water at the head of a pool takes on the characteristics of whatever type of water enters the pool. If it's a riffle, fish it as a riffle. If it's a run, fish it as a run.

The tailout of a pool, if it gathers enough speed and has a bottom cobbled enough to make the surface bumpy, has many characteristics of a riffle and should be fished with the tactics described for that type of water. Many tailouts are slick on top, more like a flat than a riffle, making it necessary to fish with finer rigging and more refined tactics. Because currents gather and accelerate on most tailouts, it's often necessary to fish from an upstream position with downstream casts, rather than a downstream position with upstream casts. If trout are feeding actively on a tailout, which usually constitutes your reason for fishing one in the first place, you're often forced to match the natural insect with a dressing that is at least a fairly close representation, in order to take trout.

Your fishing approach to the body of a pool depends on the size of the creek, stream, or river in which you find it. You'll fish a creek or small-stream pool with the same tackle and tactics you would use for all other types of water in the same size stream. When you encounter a pool in the course of a meander up a small stream, you're already rigged to fish it. If you desire to plumb the depths of a pool on a medium-size trout stream or a small river, you'll probably need to amend your terminal gear—switch from a dry fly or shallow wet or nymph to a deep nymph or even a streamer—to effectively find out what size trout might be down there. To fish the body of a pool on a big river, you'll most likely want to have a spare rod as an option. Set down the one with which you've been matching hatches and pick up one with which you can lob a heavily weighted nymph or streamer.

Tackle

The principle of matching the rod to the water makes as much sense when the subject is pools as it does with riffles and runs. You'll encounter a stream's pools as you come to them, fishing all the water types as you move along. You'll lack that handy caddy. If the rod you're carrying suits the size of the stream, you'll be able to fish the stream's pools without problems. If you're undergunned for the rest of the stream, the deficiency will show up most when you bump into a pool.

Most of us choose tackle as light as we dare when fishing for trout. That's part of the reason big pools, especially in big rivers, are not fished as often as other types of water: Sometimes we aren't armed to fish them when we bump into them, so we skip around them.

If you were to choose gear to fish just the pools along the course of any size stream, you might choose it one size heavier than you would for the rest of the water types in that same stream. A midsize trout stream that calls for a 4- or 5-weight rod to fish its riffles and runs might be fished a bit better with a rod that tosses a 6-weight line. In this age of graphite, most rods handle lines a single weight heavier without problems. It's an option to carry an extra spool wound with a line that is a size heavier than that specified for your rod, perhaps a depth-charge line of the type mentioned for fishing runs when they're blown out of shape by rain or runoff. The heavier line has the advantage of slowing the rod down a bit, which can change it from a quick one that's perfect for dry-fly fishing into a slower one that is, if not perfect, at least suitable for lobbing some of the bombs that fish best in pools.

Fortunately, the size of a bomb is again relative to the size of the stream on which it will be fished. A bomb in a tiny creek is a size 10 to 12. In a midsize trout stream, it might be a size 6 or 8. You can fish these smaller bombs with the normal run of gear you would use for the size stream on which you happen to be fishing. When most of us think of pools, we think of big, deep water in big rivers. Bombs for such water are small at size 6, are just as often size 2 to 4, and sometimes splash down tied on 1/0 or 2/0 hooks. If you're going to propel such things, your tackle must be chosen specifically to accomplish the job.

Special gear for big pools, then, should be big gear. The rod should be 9 to 9½ feet long. The line it lofts should be at least a 6-weight, but for the heaviest flies, a 7- or even 8-weight works better. This kind of equipment is not versatile. You don't want to be stuck with it when a hatch starts and you suddenly desire to make a delicate presentation with a size 20 Blue-Winged Olive dun imitation. But you also don't want to be stuck with hatch-matching equipment when you want to pitch pool-size flies.

On a two-week trip to Montana, I made the mistake of taking a light-butted, fast-action, 6-weight rod as my heavy. It was versatile, fine for fishing big drys, wet flies, and all but the largest and heaviest nymphs. But while trying to fish the banks from a moving boat, I pelted myself with so many large and heavily weighted nymphs and streamers, and failed to reach the best water so often, that I caught few trout that exceeded a pound and ended the trip frustrated.

Since that trip, I've discovered that my summer steelhead rod, a 9-foot graphite with a stiff and stout butt that normally fishes a weight-forward, 7-weight line, slows down and makes a good lobbing rod when overloaded with an 8-weight. It's an idea you should keep in mind. You might already own the perfect big rod, if only you give it a chance with the right line.

In the previous chapter, when considering tackle for fishing runs, I advised the addition to your vest of a spare reel spool carrying a sinking-tip line in a fast but not superfast sink rate. For pools of all but the largest size, you want to go at least one step deeper. There are two ways you can do it. The first is by carrying a fast-sinking, wet-head line with the first thirty feet designed to sink. The second is with a 10-foot sinking-tip line that is superfast-sinking.

The full-sinking head has two advantages: It's a little easier to cast, and all but the running line sinks. It tends to keep the fly down on the bottom with a level retrieve, rather than letting it climb upward toward the floating part of the line. The superfast sinking-tip also has two advantages: It plummets quickly, and it leaves you more floating line to control the drift and watch for indications of a strike.

I prefer the 30-foot wet-head, since I prefer whatever leaves an element of grace in my casting, and since I expect my hits in pools to be reported

Reading the largest pools is a matter of looking for the deepest water. Often you are forced to wade deep and cast long to reach it. Pablo Negri, fishing here over a pool on the Rio Petrohue in Chile, is a master at it. He also keeps well armed for it, carrying rods built for distance and lines designed for depth.

without subtlety. I also like to load it up and shoot a long cast, if needed. That is my prejudice. You should try both and see what works best for you. Whichever it is, keep it firmly seated in your mind that a slow rod works best when you're casting sinking lines and weighted flies. A fast rod will teach you to flinch every time you lean into a forward cast.

When you decide to specialize in the largest pools, with the largest trout in mind, you might choose a shooting-taper system. This consists of a thirty-foot casting head looped to a hundred feet of Amnesia or similar running line. The shooting system works best if you attach the heads with loops. Then you can carry a series of them—from slow-sinking, through fast-sinking, to a superfast depth charge line—to cover all kinds of pools. You can remove one head and replace it with another in just a minute or so. If you're forced to reel up, switch spools, and restring your line through the guides every time you want to fish a different depth, you're not likely to do it, if you're at all like me. I don't think I'm lazy, but I am modestly impatient. I'll bet you are too.

A shooting-taper system is useful on all big water and on lakes as well. When you fish the size water that calls for such gear, most of the time you'll be using a boat for transportation. It's more than just handy to have a big rod rigged and ready to reach for when you arrive at big and deep water: Big pools are big-trout water, and you want to be properly armed to fish it.

Whatever kind of sinking system you use, keep the leader short, just four to six feet, so it will not buoy the fly back up after the line has delivered it down.

Fishing the Head

The head of a pool is a funnel for food that arrives from upstream. Flies, tackle, and tactics chosen to fish the head of any pool should be dependent not on the shape of the pool, but on whatever kind of water enters it. If the head of the pool taps the foot of a riffle, it gives you a clue to the size of the aquatic insects that will be delivered down the currents and into the pool: small to medium. Choose the same kinds of searching dry flies or nymphs that you would use to explore riffled water, and you won't go far wrong at the head of a pool.

If the water at the head of a pool is shallow enough, usually not more than four or five feet, trout will often move to the surface for dry flies, even when they're not visibly feeding on top. The current tongue and the eddied water to both sides of it are excellent places to cast a dry. A size 12 or 14 Elk Hair Caddis or Deer Hair Caddis is a good choice, but you might prefer the same size Royal Wulff, Humpy, or whatever else you find to be your favorite searching dry fly. Dropping a size 16 beadhead nymph off the stern of the dry, on a couple feet of fine tippet, will never decrease your chances to find trout at the head of a pool that is fed by a shallow riffle.

Generic nymphs such as size 12 to 16 Gold-Ribbed Hare's Ears, Beadhead Fox Squirrels, or Beadhead Prince Nymphs, all at least slightly weighted, resemble a lot of the nymphs and larvae that a trout holding in the upper end of a pool sees getting washed down to it every day. To nymph the head of a pool, rig for the indicator-and-shot method and show your nymph, or pair of them, to all the potential lies along the bottom.

Trout will also move for wet flies at the heads of pools that originate in the ends of riffles. Lots of aquatic insects arrive to them in the drift, in a mix that does not reward any sort of selective feeding. The same flies and presentations that worked in the riffle upstream will continue to work as you fish your way out of the riffle and down into the head of the pool. Swing your fly, or brace of them, as slowly as you can, tossing mends as needed. Let the wet fly emerge out of the fast water that is the main thrust of the current tongue and cross the seam into the slower water on the inside of the faster current.

If the water feeding into the head of a pool is a run, deeper than a riffle, large nymphs fished deep are usually the best bet. Rig to fish two nymphs, one medium-size and one small, on the bottom with indicator and shot. Cast them upstream to the run that forms the current tongue, a few feet above the start of the pool, so they have time to sink before they reach prime lies downstream. Fish the near-side current seam along the current tongue. Probe the current tongue with a series of parallel drifts down the length of it. If the current tongue is narrow enough and your rod long enough, try to reach over and fish the current seam and perhaps the eddy, if there is one, on the far side of the tongue. This will be possible only on small streams. On larger water, you'll have to cross over in order to fish both sides of the run that enters a pool, if both sides look as if they might hold trout.

The mid-depths and surface levels at the head of a pool that is fed by a run can be productive at times, but only when some sort of insect activity has the attention of trout focused upward. This is not an uncommon condition. Whenever a hatch happens in the run upstream, a high percentage of the insects are delivered to the pool downstream. That's the major reason the head of any pool is such a favored holding lie and often a prime lie.

Fishing dry flies at the head of a pool, when the water entering is a run more than five or six feet deep, should be tied to some sort of visible feeding activity. If you see trout rising, try to determine what they're taking. Match it as nearly as you can, and present the imitation in a natural manner. If trout are not actively feeding, it's not out of line to try to drum them up with an Elk Hair Caddis or Royal Wulff. Add a beadhead nymph dropper on two or three feet of tippet. But don't expect to do wonderfully all the time. If you don't catch anything on top or toward the top, either fish the bottom or simply change the water type you're fishing. If you can't get trout to rise at the head of the pool and you don't want to rerig for nymphing, move on upstream to the riffle or run that enters it.

Fishing the Body

When you fish the body of a pool, you have three variables with which to work: fly pattern, speed of retrieve, and depth fished.

Fly patterns for the depths of pools should be nymphs or streamers and should also be large, large being defined as anything outsize for the stream on which you're fishing it—from a size 10 up to a 2/0 tied on a long-shank hook. On some rare rivers, there is a movement toward the use of streamers constructed to imitate hatchery trout, baby muskrats, and other things that might make you think it best to keep your toes out of water that contains trout that could eat them.

I fished the Metolius River one dawn with John Judy, author of *Slack Line Strategies*. John is an expert on the big bull trout known to hold in his home river. We arrived at his favorite hole at daylight and were surprised to find three people already fishing it. Paul Peterson, another bull trout expert and then the owner of a fly shop on the river, was guiding two clients who had hopes of catching a very large trout. There was only one place to stand and cast to fish the pool properly, so John and I sat under a ponderosa pine on the bank and waited a chance to take a turn. We watched Paul's clients fish for half an hour.

The pool was a strange one, a fast, five-foot-deep run with a sharp drop into deep water at the end of it. The pool itself, the prime lie for a bull trout, was about seventy feet downstream from the nearest place where you could wade into the edge of that deep and pushy run for a cast. The procedure for fishing the pool became clear as I watched Paul's first client fish. He wound up like a baseball pitcher, arced a huge fly back over his shoulder to load the line and cock the rod, then flung the fly forward about forty feet, straight across the run, where it landed with a splat. The streamer seemed about the size and weight of a dead rabbit but was actually an imitation of an escapee from the trout hatchery not far upstream.

As soon as the streamer landed, the angler frantically tossed mends into the line and quickly fed out a store of slack so that the giant fly drifted freely down the run, toward the hole. A high-density sinking line tugged it deep. When the streamer passed the end of the run and fell into the head of the pool, the fisherman lifted the rod, drew the line tight, and coaxed the fly around with a teasing swing.

To complicate matters, the fly swam straight toward a giant fallen log lodged lengthwise at the head of the pool. It was obvious that any big trout in the pool would be hunkered in the depths under that log. The idea was to retrieve the fly so it swam toward the log, then yank it upstream and out of there at the last second, so that it didn't try to swim under the log and get snagged.

After each of his clients had tried the hole for half an hour, Paul stepped in and cast for another twenty minutes using the same technique. He finally turned the pool over to John, who tried a special fly, tied out of a chunk of rabbit hide folded over and sewn together, with a tandem of 2/0 hooks snelled inside the fold. I have the fly sitting on my desk, next to a ruler, as I write this. It's more than seven inches long and fat as a frightened cat's tail.

John lobbed and retrieved this fly over the pool for fifteen minutes, but it failed to move a trout.

Finally it was my turn. The light rod I carried would not have propelled one of those flies five feet. Paul loaned me his outfit, but he changed flies before he did, tying on a tarpon fly. I don't know that the pattern has a name. It's a perfect imitation of a hatchery trout, at least on the Metolius River. It's tied on a 3/0 saltwater hook. The underwing is white bucktail, the upper wing blue FisHair. Stringers of tinsel trail along the entire length. It has a white thread head the size of a pencil end. I have Paul's fly on my desk too, also next to the ruler as I write. It's nine and a half inches long. If I got lucky and caught that fly on one of my small Coast Mountain creeks, I'd think I was having a good day.

I insisted on backcasting and forward casting this monster, in order to feel that I was fly fishing rather than pitching in a baseball game, because the act of the cast is the largest part of my own definition of fly fishing. My first cast nearly decapitated some squirrels in the ponderosa woods behind me, but it got that big fly forty feet out into the run in front of me, some distance upstream from straight across stream. I frantically stripped out running line and tossed it into the drift of the fly. When Paul, calling down directions from high on the bank behind me, told me the time was right, I stopped feeding line, lifted the rod, and teased the tarpon fly into a swimming arc toward the log.

As the line pulled tight against it, the fly felt heavy. I said to Paul over my shoulder, "This thing is so damned big it feels like you have a fish on when it comes around in the current." I turned back to my fishing, stripped in some line, and tried to give the fly a couple more teasing tugs to animate it. That monstrous saltwater streamer kept feeling heavier and heavier. The rod bent lower and lower. When its tip got down near the water, I reared back on it as if to ask it a question: "What the hell's going on out there?"

The answer was a strong tug back.

For a minute, the fish or the fly or whatever it was felt loggy, and I assumed that's what I'd hooked: the log. Then it defined itself by rushing off downstream toward the lower end of the pool, turning and swimming under that log. I brutalized it out of there, forced it upstream out of the pool and into the run, and held on while it tired itself out thrashing around in the fast water.

The trout did not put up a brilliant fight, but neither did I. A brown or rainbow in the same kind of heavy water likely would have torn me up. But the bull trout wasted its strength on the current, then let itself be led out to the edge of the fast water. It slid over my hand without much more struggle. I hefted it out of the water for some photos, then slipped out the 3/0 hook and aimed the trout back toward its home in the pool under the log. It needed to grow. It was a fairly small one, only six pounds.

I looked at the fly, shook my head in wonder at the size and success of it, and handed the rod back to Paul. "I'd be a fool to take another cast," I told him.

When fishing the body of a pool, you should select your fly based on two things: the size of the trout you hope to catch and the size of the bites you think that trout most commonly eats. That bull trout on the Metolius River was a few hundred yards downstream from the outlet of a fish hatchery that grew stockers for other waters. What the bull trout ate came sneaking out an effluent pipe.

Most big trout are predaceous, though they're not all piscivorous, and few are so fierce as bull trout. They don't all feed on hatchery trout or baby muskrats, but they all like big bites.

If the trout you hope to catch from the body of a big pool weighs two to four pounds or more, then you could expect it to feed most often on sculpins, crayfish, and small baitfish such as minnows and chubs. Though it undoubtedly likes them larger when it can capture them, most of its meals will be around two to three inches long. These are imitated well on hook sizes 2 and 4, long shank. You can go a lot bigger, but the guides I have talked to and fished with in Montana feel that flies tied on size 2 hooks will interest the largest trout, such flies are reasonably easy to cast, and larger flies increase the difficulty of casting without increasing the numbers or size of the catch.

The most popular flies for big trout are sculpin imitations such as Muddlers and Spuddlers, Woolly Buggers in black and olive, streamers such as the Spruce, and big ugly nymphs like the Bitch Creek or Yuk Bug. My favorites are black and olive Bunny Leeches, tied with tungsten beadheads and a few extra turns of nonlead wire on the hook shank. Many other flies might work as well. Zonkers are favorites in lots of areas. Strip Leeches work well for large trout. You should always use what has worked in your own past.

Two conflicting theories exist about whether flies for pools should be weighted or unweighted. The first is that the line should be used to get the fly down, while the fly should be left unweighted so it moves more naturally in the water. The second says the fly should be heavily weighted so it gets down and bumps the very bottom no matter what kind of line is used. Everybody must choose what suits best. Some people don't like to cast fast-sinking lines, preferring to cast heavily weighted flies. Others don't like to cast weighted flies, preferring to cast dense lines.

My own practice treads middle ground. I weight my large flies moderately, with fifteen to twenty-five turns of nonlead wire the diameter of the hook shank. This is enough weight to get them under the surface quickly. They work well when cast tight to banks along the edge water of long runs from a boat. An unweighted fly in this situation might not sink at all and would catch few trout when retrieved so shallow it left a wake.

A fly weighted modestly gets down deep and fast when fished with a sinking line. But it still has some life, some movement of its own, when fished on the bottom. If the fly were weighted like a stone, it might also fish like one. I don't weight big flies heavily very often, because that narrows my options, reduces the ways I can fish them.

Once you've chosen your fly pattern, the speed of retrieve is the second consideration. There are essentially three speeds: fast, medium, and none. The first is often the best, because it imitates the movement of a lot of the trout foods that big trout chase. A fast retrieve looks like something trying to escape, which is what trout expect things to do when they prepare to consume them.

A medium-speed stripping retrieve is also excellent, because it represents the stop-and-go movement with which most natural foods swim most of the time. Your fly might look like a natural baitfish or other food form out for a brisk stroll, not aware that it's about to be surprised.

No retrieve at all works wherever the current in the depths of a pool is adequate to give the fly some action, which turns out to be in most pools. This is why no retrieve might actually be the retrieve you should use most often. If the pool has a fair flow, it sometimes works best to cast the fly out, let it get right down to the bottom, then allow it to tumble along at the request of the current. This is the way a lot of food is delivered to trout, and though there is no indication that they feed much on baitfish that are already dead, they definitely kill cripples. Chances are the current will give your fly enough movement to make it look alive, though perhaps stunned. If the current is at all quick, it will swim the imitation just as a natural might move along the same bit of bottom water.

One retrieve might work best one time, but the next time it will fail to move trout at all and another retrieve will work better. You have to experiment until you find the one the trout want at the moment. Use alternate retrieve speeds on successive casts. Sometimes you can try all three retrieves on the same cast: Let the fly sink to the bottom and dead drift a ways, then pick it up by stopping the rod and tightening the line, tease it around a part of its arc on a medium-speed retrieve, and finally bring it back to you with fast strips that swim it through the water at a trot.

I'm far from a conventional angler here, but it's hot on Hounds Creek in Montana in midsummer. It's a bit out of focus, but a run slows down and deepens into the body of a pool right along the bank behind me. A big rabbit strip streamer cast there and allowed to sink for some time before it was retrieved deep brought this brown trout up after it. The trout was in sight and seemed about to turn away when greed got the better of it, and it jumped the streamer. Had I come out of those depths and beheld me in that outfit, I'd have turned and fled for cover in fright and not gotten caught.

The final factor is depth. If you want to catch big trout, then think *bottom*, especially in big pools, unless the trout are actively feeding. When fishing the body of a pool, forget about the mid-depths and surface. If trout are out cruising, they'll not be doing it in the middle of deep water. Instead, they'll move to the head, tailout, or shallow edges of the pool, where stream life is richer. For the purposes of reading water, consider all but the bottom of the body of a deep pool to be a wasteland. If you fish anywhere except where trout can at least see and have a chance to chase your fly from their stations near the bottom, you'll usually be wasting your time.

It's not always easy to accomplish getting your fly right to the bottom of a deep pool. You can think you're doing it while missing by several feet, retrieving your fly so high above the bottom that trout either fail to notice it or refuse to move for it. I'm not good at it myself. I haven't set up lunker hunting for big trout in big water as much of a goal in life. Many of them have come to me simply because I'm out there fishing more than I should be. Most of those have been taken in association with the bottom, in one way or another.

It's best to take up a casting position at the side of the pool, if the shape of the pool allows it, where you can quarter your casts slightly upstream. Let the fly and line sink as they drift downstream. Begin your retrieve as the fly swings straight across from you or slightly downstream below you. This gives you the best control. But the pool often dictates the way you can fish it.

If you can wade in only at the head of the pool, and the entering current has much force, you'll need a faster-sinking line than you would in slower water, since the current will buoy the line and fly upward. If you're forced to cast to the body of the pool from a position at the tailout, your line might need to be a slower sinker, because the current coming toward you will let the line and fly sink more swiftly. In this case, you'll want to speed up your retrieve; you have to activate the fly in relation to the current, not the bottom.

If the body of a pool has little or no current, you can fish it from any casting position the shape of the pool allows and achieve the same result. You want your fly to swim within a foot of the bottom. You should be feeling it touch now and then, or you're not fishing deep enough. Expect to lose an occasional fly if you're fishing a large pool right.

It's dark down there, and the trout you're after might not see the fly unless it comes very near. Don't just take a cast or two and assume you've covered the pool. Set up a disciplined pattern that covers all the deep water. If you *know* big trout are down there, your casting pattern should cover all the water two or three times. When you're after one big trout, rather than a bunch of small ones, you need all the patience you can gather. It doesn't happen often that you catch a six-pound trout on the first cast. It's never happened to me again.

The body of a pool is a bit like a pond or lake. You must experiment with the three variables—fly pattern, speed of retrieve, and depth fished—until you find the right combination. The fly pattern should be one in which you've developed confidence. The depth should be near the bottom. The right retrieve might be about three times as fast as most people think would frighten a fastidious trout, but most times a dead slow retrieve works best.

It never hurts to try something different every few casts, just to test the waters. Change flies, fish different depths, and vary the speed of your retrieve until you find what works.

An Accidental Trout

When I spoke at an annual conclave for the Hickory, North Carolina, Trout Unlimited chapter some years ago, James Fortner (shown above) and Carl Freeman invited me to go fishing with them in the nearby part of the Appalachian Mountains. They didn't blindfold me, but the simple trick of not telling me the name of the stream we fished has so far kept me from naming it. They told me to expect a beautiful mountain stream, with lots of plunge pools typical of stair-step water and no trout larger than ten to twelve inches. The stream was more beautiful than they'd described it. Most of the trout we caught were indeed ten- to twelve-inchers. I used a dry fly, a size 12 Deer Hair Caddis, to suspend a size 14 Beadhead Hare's Ear on a three-foot tippet. Takes were about equally distributed between the two flies.

In one of the larger pools on that nameless stream, when my indicator dry fly dipped under, I raised the rod to set the hook and came up against what felt like weight attached to the dangled nymph. The trout fought hard but was reluctant to leave its

large pool, so I eventually brought it to hand on the 5X tippet I'd used for the dropper. Carl (left) and I couldn't decide whether to lie and say it was six pounds or lie bigger and say it was eight. It was at least four and exceeded all our expectations for that pretty stream's pools by about three pounds. It had obviously been making its living for a long time by taking up station in the biggest pool it could find and eating the smaller trout we'd been catching. I have no idea to this day why it decided to nibble on my little nymph.

Fishing the Tailout

The tailout of a pool is less complicated than its head or body. It should be fished as you would a riffle or run, keying your fly selection, tackle, and tactics to the hatch that draws trout into active feeding there. Most of the time, those insects will be relatively small. Often you'll be forced to match a hatch, since the surface of a tailout tends to be smooth.

The three-layered approach—bottom, mid-depths, and top—becomes important again when tackling tailouts. Trout often focus their feeding on drift close to the bottom. They also feed regularly on aquatic insects in the vital transition from bottom to top, in the mid-depths zone, during the early stages of a hatch. When the greatest concentration of food is on the surface, trout hang high in the water and feed willingly on the top, even at high risk of predation from overhead.

Fishing nymphs along the bottom of a tailout calls for upstream tactics similar to those used in a riffle, with weighted flies, a leader the length of the rod, and a floating line. A strike indicator placed about twice the depth of the water above the fly will increase the number of takes you detect. Split shot or putty weight on the leader, between two nymphs or very near the single nymph if you're rigged with just one, will help get you down where you need to be.

Probing the mid-depths of a tailout with a wet fly can be one of the most effective tactics available, but it's one that is rarely used today. I try the downstream wet-fly swing by choice when trout feed visibly in the thin water of a tailout. Even if I see adult insects on the water, my success is usually better in these situations when I select a wet fly the color and size of the natural and present it just under the surface. This works especially well on broad tailouts, where it's difficult to wade into position for dry-fly presentations.

Fishing dry flies on tailouts works well, but certain aspects of tailouts make it difficult. First, the water accelerates as it narrows on the tailout. Casting from a position downstream subjects the fly to drag after just a few inches to a couple feet of free float. Second, it's often difficult to approach from upstream, since that's the deep part of the pool, and it's hard to cast well while you're treading water. In many tailouts, however, you can wade into position to cast laterally and fish with reach casts straight across stream, or you can position at an angle upstream and fish with downstream wiggle casts that place the line and leader on the water with slack, for a drag-free dry-fly drift.

This difficulty getting into position for an effective presentation with a dry fly is the major reason I have most of my success on tailouts with wet flies. I can cast long, from a position at the outside edge of the tailout, and swim the fly slowly across the lip, where the trout feed. Drag would kill any hope for success with a dry fly. With the wet, the swimming fly looks like an emerging insect on its way to the surface. But it mustn't race. If a wet fly gets going too fast, trout consult their books on aquatic entomology, find nothing in nature that moves that way, and refuse to eat it.

When large trout move out of the depths to forage on a tailout, usually at dawn or dusk, the same wet-fly swing works well, but with a larger streamer. The same Muddlers and Woolly Buggers and Bunny Leeches that worked in the body of the pool will work on the tailout when large trout are feeding there. Fish these streamers on lines, often floaters, that keep them just under the surface. In the shallow water at the tail of a pool, you don't have to worry about getting your streamer down to the bottom. Trout will see it at any level.

If a large trout is actively hunting a tailout and spots your large fly, it will most often wallop it. Sometimes you'll see an arrow drawn in the shallow water, pointing straight at your streamer. Don't panic and set the hook until you feel something solid out there.

8

The Challenge of Fishing Flats

Flats are shallow water, from one to four feet deep, with gentle gradients that produce slow flows and a smooth surface. If the source is a spring creek or tailwater, a flat usually has rooted vegetation, myriads of small aquatic insects, and prolific hatches, and it requires fishing with delicate gear, imitative flies, and drag-free presentations. Because of their gentle currents, plant beds for overhead protection, and abundance of food, stabilized flats hold many trout, often very large ones.

Tilt the gradient of a flat down a bit and it would flow faster, no aquatic vegetation could take root, and the bottom would become a collection of coarser material: cobble and rock. The surface would reflect the roughness of the bottom, and you would have a riffle or run, depending on its gradient, depth, and bottom material.

Freestone streams have flats as well, but because of winter spates, their bottoms are cobble, which tends to be very uniform. If the composition of the bottom rock is big enough and varied enough to break the current in places, the current is slow enough that trout don't have to fight it continually, and the water is deep enough or has enough hiding places for trout to avoid overhead predation, then it will hold trout. Almost by definition, however, if it meets all those criteria, it qualifies as a riffle, and we already know that riffles hold lots of trout.

A typical freestone flat has cobble too uniform, current too swift, and depth too shallow to be anything but empty water. Perhaps that is too restrictive a classification. Many freestone flats are very productive for aquatic insect life. Small trout and whitefish find satisfying lies on and just above the bottom cobble. They are fed well enough that they'll risk predation to take up territories and stations even on some shallow freestone flats. But big trout won't.

191

A flat is a reach of stream where the water slows, as it does in a pool. But instead of nearly stopping, it retains its defined current. Instead of deepening, it spreads out, with an almost even depth from side to side. Because the water is slow, the finest material is allowed to settle to the bottom. If the water source is stable, the bottom will not scour.

Many flats have bottoms of fine cobble and pebbles. Most have at least some sediment and sand mixed in with the stones. This has two results: First, the smooth bottom is reflected as a glassy surface; second, the bottom makes fine soil for rooted aquatic vegetation.

Flats are shallow. Sunlight strikes through to the bottom, and photosynthetic growth is at least as rich as in a riffle. But the current of a flat is gentle. In a riffle, vegetation is constantly scoured out during high flows. Plants don't have a chance to get a grip on the bottom. In flats, plant beds are so common that they are part of the definition of the water type. Productive flats are almost always set apart from unproductive flats by the growth of rooted vegetation or attached algae. Such growth indicates current speeds that are extremely suitable to trout. The growth offers a richness of insect life. Plants also offer protection from predators, converting lies that would otherwise be used only for feeding into prime lies, where trout can be found at all times.

The Railroad Ranch section of the Henrys Fork of the Snake River, in Idaho, is the ultimate flat. It has two major sources, one a spring creek, the other a tailwater controlled by releases from Island Park Reservoir. Its flows are steady, and the long, broad flats are not subject to scour. The entire bottom is bearded with a growth of plants that become longer as the season gets older.

The Ranch reach is about three miles long. In the upper mile, most of the river is about a hundred yards wide and can be waded across with some caution. It averages three feet deep but has some surprise trenches a foot or two deeper. The current is steady and firm, but you can easily wade against it. In the middle mile or so of the reach, the river spreads out farther, deepening in places so that wading requires extreme caution, and the current slows to almost nil. Farther down, there is a brief stretch of rapids, then another half mile of flat before the river flows under the highway and out of the Ranch section.

The water is spring creek clear. Those great plant beds are hotbeds for aquatic insect growth. Hatches are heavy, sometimes so heavy you want to wade to shore in disgust because it seems that the chance a trout might select your artificial among hundreds or even thousands of naturals would

The defining feature of a productive tailwater or spring creek flat is rooted vegetation, such as these trailing plant beds on Montana's Bighorn River. The plants are able to get hold of the bottom only where stabilized flows allow the deposition of finer material—gravel, sand, and even silt—and can remain in place only in waters that lack scouring high flows in winter and spring. Streams with spate sources have flats, but they do not allow fine bottom material to settle, and they have seasonal storms and snowmelt that remove any plants that might attempt to take root.

Plants provide millions, perhaps billions, of microniches for aquatic insects and crustaceans. These prod trout growth and also sponsor the sorts of hatches that make it possible to read flats by looking for the sorts of rises visible in the photo just upstream from Jim, who is focused on a pod of trout rising in front of him.

be remote. Sometimes it is remote, and it feels smarter to mismatch the hatch in hopes that a trout will take your fly because it's different. But most of the time, trout can be caught with the right imitation if it's fished with the proper presentation.

Rainbow trout work in pods across Henrys Fork flats. They move slowly upstream, feeding as they go, on whatever insect is coming off at the moment. After a hundred feet or so, which might take fifteen minutes or two hours, they seem to drop out of sight for a few minutes, then materialize again, rising at the lower end of their favored beat to begin working upstream once more.

Fishermen spread out to work on these pods of trout. Through the compression of a telephoto lens, it can be made to look as if they're so close together that their backcasts would tangle in the air. But they're a comfortable distance apart, and each has his own pod of trout to work over. It's one of the few places in the world where I have fished in what I would normally call a crowd without feeling crowded. The single time I saw a problem was when a guide in a high-prowed drift boat fished a client down through the Ranch flats during a green drake hatch. The boat was unnecessary. The guide waded at the bow, slowly pushing the boat downstream, while his client stood in the boat and thrashed the water. The client caught no trout.

He had no chance of catching any trout the way he went about it. His guide was probably trying to hurry him downstream to the riffles.

We tend to associate flats with spring creeks and tailwaters. That is because *productive* flats are found most often on those types of water. But flats are a result of a stream's gradient, not its source. Most freestone streams have occasional reaches that are just as flat as those with stabilized flows. Some of them are just as difficult to fish. My friends and I fish the Deschutes and Willamette rivers in Oregon several times each year. Our favorite drifts on each river have reaches that we nicknamed the Henrys Fork.

The Henrys Fork of the Deschutes is two hundred yards of side current. It is fifty yards wide and three to four feet deep in toward shore, too deep to wade twenty feet out from the bank. The flat has a sandy bottom with lots of trailing plant beds, a strong but even current, and a flat surface. Trout rise consistently to hatches, and the tactics that take them are the same that we use on spring creek flats. We fish that twenty feet next to the bank and do very well there. It's not legal to fish from a boat on the Deschutes. Trout that rise out of our reach on this flat always look at least twice the size of the fourteen- to sixteen-inchers we catch in the part of it that we're able to wade.

Some famous flats, such as this one on Idaho's Henrys Fork of the Snake River, seem far too well attended, especially during the bloom of mule's ears flowers, which signals the green drake hatch. Nevertheless, it's usually possible to find a pod of rising trout and have them all to yourself. Reading the water in this case is as simple, but also as complex, as watching for rises. Never fish such water without a pair of small binoculars, so you can be the first to notice sipping rises around rooted vegetation beds in midstream or tucked in along the grassy banks and wade into your position before the trout get surrounded.

The Henrys Fork of the Willamette is the broad tailout of a pool that is a quarter mile long and more than a hundred yards wide. The water rises up to four to five feet deep, maintaining this depth for the last hundred yards of the pool. It has a rocky bottom and little attached vegetation. Its current is easy and its surface smooth.

Trout constantly rise across this flat tailout in spring, summer, and fall, feeding selectively on tiny mayflies, caddisflies, and stoneflies. The water is too deep to wade safely except on the tailout. We drift down it in boats, anchor, fish what water we can reach, then raise the anchor and lower the boat a few feet down the current before dropping the anchor again. Trout sometimes rise right at the ends of our oars, but it can still be exceedingly difficult to determine what stage of what insect they're taking and find a fly that imitates it successfully. I find the fishing easier to solve on the real Henrys Fork than I do on what we call the Henrys Fork of the Willamette.

The flat stretch of water that we call the Henrys Fork of the Willamette is wadable, has a glassy surface, and is often as difficult to fish as its namesake. We fish spring and summer march brown, blue-winged olive, and little sister sedge hatches there. Rainbow and cutthroat trout, and sometimes whitefish as well, rise to take them selectively. Again, we read the water by looking for rises, but it's our knowledge of the river that lets us know we need to float fast and cover some water in the morning, so that we can anchor up on the flat in time to eat lunch and scan the water until the hatch starts.

It's an aside, but it's also easily overlooked, that knowledge of hatches is a form of reading the water. If you know where and when to meet trout rising to a particular insect, then in the broadest sense that is reading the creek, stream, or river you are setting out to fish.

Freestone streams have many unproductive rocky flats that border on riffles. These are generally empty water, and you should learn to recognize them so you don't spend a lot of time fishing them. They're usually long and broad, a foot or at most two feet deep, with bottoms of such uniform cobble that trout of any size find very few places to shelter. The surface of such a flat might be smooth or have a slight chop, depending on its speed and the size of the cobble on its bottom. Whether the flat is smooth or rough, it will almost always be unfeatured, and water without features, in the absence of aquatic vegetation, is usually water without trout.

These kinds of rocky freestone flats either are too fast to allow the growth of rooted plants or suffer scour during winter storms and spring runoff. They do have excellent populations of aquatic insects, and trout might move up into them to feed if a hatch is heavy. For the most part, though, trout avoid such flats because they have to fight the current and risk predation, which makes this kind of water a poor place to be in terms of that binding energy equation.

NEEDS OF TROUT

Flats meet the needs of trout, or fail to meet them, in predictable ways. Because freestone flats tend not to meet the need for shelter from currents, and can usually be both defined and fished as riffles when they do, the following discussion will focus on spring creek and tailwater flats. Never forget, however, that some stretches of freestone streams can take on meadow stream characteristics. In any creek, stream, or river, a flat that is somewhat shallow, has a level and fairly smooth surface, and supports an abundance of aquatic insect life will fall into this category. Its water should be read, and its trout pursued, in the same way you would scout and then fish a flat in water with a stabilized source.

Shelter from Currents

The need for shelter from currents is met easily on most flats. Currents are gentle, and trout need little protection from them. Extensive plant beds that typify flats create a pattern of channels, eddies, and still pockets that give trout lots of places to take shelter. Most trout on flats find their comfort in plant bed cover.

A few obstructions to the current can be found on any flat: boulders, trenches, ledges, stranded logs, shopping carts, car parts, and anything else that becomes lodged in a current too subtle to wash it away. All these structures break the current and form protective lies. Trout tend to hold in relationship to some sort of obstruction, most often a plant bed, when they're not feeding. When a hatch happens, as many insects might emerge around these kinds of cover as elsewhere, so trout tend to remain on their lies, in their territories if not riveted to their stations, unless those lies fail to feed them.

Protection from Predators

Protection from predators is the least common denominator on flats—the need most poorly met. The water is generally shallow, as clear as trout stream water gets, and smooth on top. There is little to impede the sight or

The same plant beds that feed trout also break whatever currents they might find disturbing and give them concealment from overhead predation, at least when they're not exposing themselves by rising to feed on a hatch of insects. These channels in the vegetation on Spring Creek in Pennsylvania, near State College, form perfect lies for trout, but they're difficult to fish when trout are holding along the bottom. You almost have to spot one before you can deliver a fly to it accurately enough to have a chance at the trout.

The slightly cloudy appearance of the water is not from wading or from runoff or rain. It is the natural color of the water from leachings through the limestone formations that form the bed of the stream and add the chemicals that enrich the stream. That slight opaqueness is to your advantage. If the water were clear, trout could see your approach and detect the fraudulence of your flies even more accurately than they already do.

hinder the attack of an osprey from overhead. But the same plants and occasional obstructions that provide shelter from minimal currents also provide protection from predators when trout hold their stations. A station that did not provide cover or concealment from overhead, or that was not deep enough to keep an osprey from penetrating to the level of a holding trout, would not be a station and would not hold a trout. Trout are most vulnerable to overhead predation when they move off their stations to feed.

When trout feed on hatches on flats, they must move up in the water column and usually out into the open, exposing themselves. Trout are almost always willing to risk danger when a food form is abundant and available, even to the extent of losing a tithe of their number. During the duration of the long trout season, some sort of aquatic insect or terrestrial usually makes itself available to trout at one time or another in the course of a day on a flat. Then trout are drawn out of their hides to feed. That is what draws us to flats.

When trout expose themselves on flats, they are as wary as trout get.

Trout Foods

The need for food is met on flats by the great abundance of aquatic insects and crustaceans that thrive in plant beds. The ever-observant Charles Brooks pointed out the reason: *vertical use*. Aquatic insects are not confined to the substrate, where they find niches in and on the rocks of the bottom. Instead, they use the entire volume of the water column into which plants grow. Plants provide even more microniches for insects per cubic foot than the cobble of rich riffles. Insects, like trout, are most abundant where they find the most food. Plant beds provide much richer pastures for insects than do bottom stones. A flat with rooted plants can furnish astonishing numbers of aquatic insects, crustaceans, midges, aquatic worms, and other items that nourish trout.

The most important aquatic insects on most flats are small mayflies, though that level of importance might be related to their propensity to hatch en masse as much as it is to their sheer numbers. Blue-winged olive (*Baetis*) nymphs are widespread, all across the continent, and thick in plant beds. If you run an aquarium net through trailing tendrils, it will often come up squirming with them. They're tiny swimmers; if you put them in water, or just hold them in your cupped palm and add water, you'll see that they swim like little minnows. Flats trout feed opportunistically on these nymphs almost continually, which is why a size 18 to 22 Flashback Pheasant Tail works so well as the trailing pattern in any two-nymph tandem, even over big trout.

Blue-winged olive duns are the defining hatch on many spring creek and tailwater flats. I've seen them come off in such numbers on a Missouri River flat, on a rainy day in September, that the surface in the central currents turned a smoky gray. I was drifting the river in my pram and was mystified by it; it seemed like some sort of mist. I rowed into the mist and peered over the side of the boat. Only by looking closely could I tell that the currents were covered with tiny size 24 blue-winged olive duns.

That discovery didn't help my fishing much. Trout were selective to them, and though I had imitations, mine got lost among so many. I didn't begin to do well until I tied on a monstrous and gaudy mismatch: a size 22

This is the key to reading water on flats: a rising trout. It would be wise to observe that in this instance, however, the trout, though breaking the surface with its back and tail, has not left a bubble of air in its rise ring and therefore almost certainly took something just beneath, and not in, the surface. It's very important to notice such things, especially in the presence of all those insects floating on the surface. If the trout are feeding subsurface, you can use a dry fly all day and catch little but frustration. Reading riseforms is an adjunct to reading trout water.

Klinkhamer with a pale orange wing post. For some reason, a few trout separated it out from the naturals and took it, and I could tell when it happened. One of those trout weighed three pounds, and a bigger one made a long dash at the sting of the hook, taking out so much line that the weight of the line gave the heavy trout a lever against which it broke my 6X tippet.

Tricos (*Tricorythodes*) in size 20 to 24 come off in heavy hatches in late summer and fall on flats east, west, and in some places in between. Morning sun striking millions of shiny wings in their spinner flights can turn the air over streams and rivers into tall columns of bright reflective fog. When Trico spinners fall to flats, trout come up and feed almost frantically on them.

Browns and rainbows hold in pods, poke their snouts out of water, and inhale insect after insect, rarely dropping more than two or three inches below the surface between takes. It's so tough to hook them that many folks tie doubles on an outsize hook—size 16—and hope to entice a trout that is twice as greedy as all the others. It works. Sometimes mismatching Tricos with a size 16 or 18 olive Spent Partridge Caddis works as well. Most times, it's best to fish an imitation and try to time your casts so your fly arrives just as a trout is ready to rise up and inhale another natural. Tiny soft-hackles, fished just beneath the surface, are another solution.

A few larger mayflies also inhabit plant beds in scattered flats. Western green drakes (*Drunella*) are an example. These size 12 insects are abundant on the Henrys Fork, but their hatches are elusive, subject to whims of wind and weather. Pale morning duns and sulphurs (*Ephemerella*) are sizes 16 and 18 and much more widespread. Their hatches go on for weeks and last for hours each day. They are abundant, but they don't blanket the water like some of the smaller mayflies. If you get into a hatch of them, they're easier to solve, and you can usually fish their imitations on 5X tippets, which gives you a better chance to land the size trout you'll hook on many flats.

Overlapping hatches are common on flats, with two or even three mayfly species emerging at the same time. All the trout might key on a particular species, ignoring the others. Just as often, individual trout might prefer different species, forcing you to pattern a specific trout and determine which insect it prefers. To complicate matters, two stages of the same species, nymph and emerger or emerger and dun, might be present. While it may appear to the casual observer that a trout is taking duns from the surface, in reality it might be feeding on emergers just beneath the surface film.

The most common scenario, in my experience, is a hatch of size 18 or 20 blue-winged olive duns masked by size 16 pale morning duns. Without binoculars to spy right among the floating insects and poking noses, it's difficult to notice that trout almost invariably prefer the smaller insect.

Flats are a challenge.

Caddis can find the microniches of flats perfect habitat and sometimes become almost as abundant as mayflies. Certainly when they're in the middle of an emergence, and mayflies are not, they're much more important.

Net spinners, such as the spotted sedges (*Hydropsyche*) and little sister sedges (*Cheumatopsyche*), are most abundant in spring creeks and top-release tailwaters. They run from size 16 to 18, and you need accurate imitations when lots of the adults are afloat on flat water at the same time.

Case-building caddis are important on many flats. Long-horn sedges (*Oecetis*) are found on scattered flats, but where they're abundant, such as on the Missouri River tailwater, they're very important. These also are small, size 16 to 18.

Grannoms (*Brachycentrus*), also called black caddis, are most common in bottom-release tailwaters, where the water downstream from the dam is usually colder because it comes from the depths of the reservoir. Grannom pupae and adults are sizes 16 and 18; trout might focus on one or the other. Related weedy-water sedges (*Amiocentrus*) are smaller, sizes 18 and 20, and less dark. Their hatches frequently overlap those of the larger black caddis, and trout often prefer them, sipping the smaller insects among a fleet of larger ones. You need to notice such things. Reading hatches, on flats, becomes a key part of learning to read water.

Midges on lakes and ponds can attain large sizes, 8s and 10s; tie your imitations on long-shank hooks. On spring creek and tailwater flats, the moving water types where midges are by far most important, they're almost always tiny. A size 16 midge would be a big one on a flat. Size 20 to 24 is closer to average. In almost all circumstances, you need only a few dressings in the right sizes and a few colors to imitate midge pupae. Examples include Thread Midges in black, red, and olive. Forget midge larvae and adults; trout take lots of them but not often selectively as they do the pupae.

Terrestrial insects become important on flats at times, almost always in the warm days of summer and early fall. They're most likely to be taken selectively along the edges, near grassy or brushy banks, where they fall to the water and trout sometimes line up to greet their arrival. Tiny ants and small beetles are the predominant terrestrial forms along the edges of flats. These can be difficult to notice unless you suspend an aquarium net half in and half out of the surface film to strain the currents. Then all you have to do is calm yourself down, find among your fly boxes the minute pattern that matches the terrestrial you've discovered, and present that imitation correctly to the fussy trout.

Grasshoppers are less frustrating. They often tumble to the surface of a flat, again along the edges. On windy days when lots of them suffer the same fate, trout get excited. The resulting dry-fly fishing can be a succession of detonations. It doesn't happen often, but it's a lot of fun when it does.

Crustaceans such as scuds and aquatic sow bugs, also called cress bugs, are prevalent in waters with a high alkaline content and dense plant or algal growth. Trout nose into the greenery, flushing crustaceans like quail, gunning them down in channels between tendrils of plants. This kind of feeding creates some of the most difficult fishing. Trout work among the plants or right along the bottom. They take only the tiny naturals, and they're spooky feeding in those clear currents. You need to cast small imitations, well weighted and often headed with tungsten beads, suspended on long, fine tippets beneath very small indicators. Your approach and presentation must be more cautious than they would be if the same trout were rising, feeding on the surface.

When you get it right, the rewards can hang heavy in your hand, if they don't get caught up in the plants and break off.

It's critical to remember that the most important trout food, insect or crustacean, on any flat is the one that trout are focused on at any given moment. Usually that will be the one that is most active and available, perhaps emerging or just shifting home base on the biotic drift. The detail of these notes on the food forms of flats is relative to the importance of hatches on flats: If trout are going to feed selectively, they're most likely to do it on a spring creek or tailwater flat. If you're going to take refuge in detailed and perhaps regional hatch guides, you'll most likely be driven to them by situations found on flats. I recommend that you refer to those types of guides if you fish flats a lot, because much of reading water on flats is finding feeding trout and matching what they're taking.

Temperature and Oxygen

Temperatures rise more rapidly on flats, which are slower and more open to the sun, than they do on faster, more likely shaded, and more often broken parts of any stream. Trout might be forced into torpor, and perhaps into those other water types, sooner on flats than they are in riffles and runs. But a high percentage of fishable flats are found on spring creeks. These arise from the ground at constant temperatures year-round and carry their coolness downstream for many miles. It's rare that temperatures on spring creek flats reach levels that are distressing to trout.

Tailwaters are the locus of just as many, or even more, productive flats. Their water is delivered at the temperature of the reservoir upstream. If it's a top-release tailwater, it will arrive warmer in the heat of summer. Because flats are broad and shallow, such tailwaters might have only a few miles of trout water, downstream from the dam, before they revert to the warmwater fisheries they probably were before the dam was installed on the river system.

Bottom-release waters arrive much colder, shifting some of the types of insects found there and extending the distance the water flows before it inevitably warms in summer, temperatures become too high and oxygen levels too low for trout, and the river gives itself over to warm-water species. Some tailwaters provide only a few miles of good trout fishing. Others provide many miles. It all depends on the type of release and the climate of the region through which the river flows.

Each tailwater offers a different set of temperature and oxygen regimes, and each has a different distance that offers good trout fishing. It's easy to consult local fly shops for the borders, though if you extend those boundaries downstream, you will on rare occasion catch a very big trout, usually a brown, that remains almost dormant on its station at the bottom during the heat of the day, then takes walks in the cool cover of darkness to get some exercise and murder luckless trash fish.

TROUT LIES

Most of my reading of trout water on flats consists of watching the water very carefully, sometimes with binoculars, for rising trout, because fishing hatches over rising and selective trout is most of what I travel to streams with productive flats in order to do. I recognize it as a defect in my personality: I don't drive or fly all the way to the Bighorn or Missouri River to fish the bottom with nymphs, though I'll do that happily if no trout are rising. But I read the water first by searching it for rises.

A second part of my reading of water on flats consists of peering into it, trying to find trout holding on their stations in their territories or else nosing into plant beds, feeding on nymphs, larvae, and crustaceans deeper in the water column or even on the bottom. When I'm able to spot them and watch what they're doing, I know exactly where to fish for them and what flies and methods to use.

Reading the water to ascertain the most likely lies of nonfeeding trout is an important part of any flats strategy. But it can be secondary or even tertiary to spotting the trout or their rises.

Finding holding water on flats, when trout are not rising or otherwise visible, is one of the easiest assignments in reading water. Because of their uniform flows, flats usually carry their depths right to the edges and have high banks from which you can scan the water for signs of likely lies. Wear polarizing sunglasses.

Plant Beds

The structure of a plant bed forms the most obvious lie on a flat. Trout hold wherever plants are deep enough to give them overhead protection or in shallower beds if they can find concealment in them. Beds that are fissured by minor crevices and canyons offer the most potential protection. They're also among the most difficult places to hook trout. You have to get close in order to find an angle of presentation that keeps your line, leader, and fly

Plants are structure to trout on vegetated flats. They hold in fissures among and alongside them, as if they were boulders or lodged logs. If you are able to spot these trout, it means the water is quite clear, and they're just as able to spot you or notice any mistakes you make in your presentation. Finding them is important, because if you don't, it's nearly impossible to simply read the water and fish for them as you would most freestone situations. But you aren't halfway to catching a trout when you've found it, even if you've seen it before it has seen you. It's still necessary to refine your rigging and present your fly in a way that shows the trout nothing of your line in the air or leader on the water.

clear of the plants while your dry fly floats over the narrow slot or, more difficult, your nymph sinks down into it. Trout in such water, which is usually clear and somewhat shallow on flats, are extremely wary. Getting close to them without spooking them, or at least alerting them to your threatening presence, is not nearly as easy as it is in a riffle or run of similar depth.

Obstructions

Any visible shelter or protection is an obvious place to find trout, especially when they have not moved off their stations to feed elsewhere. Boulders, either protruding or revealed by gentle boils on the surface, are excellent prime lies on flats. It's common to find logs wedged in flats; the current is not always strong enough to dislodge them.

Fallen logs form structure on flats with stabilized spring origins, such as Fall River in central Oregon, whereas they would quickly be swept away on streams subject to spate flows. They break the currents and provide bomb shelter–type hiding places that allow fish to escape predators, and if they've been in the stream long enough, as these pines have, they become habitat for insects and thereby feed trout holding in the currents downstream.

All the currents broken by the limbs and logs on this flat have potential to hold trout, but the prime lies are tucked tight up against the log and also in the line of drift from the log down along the right bank, which is probably undercut and therefore the likeliest lie for a large trout on this flat.

On many narrow spring creek flats, jackstraw tangles of fallen trees lie anchored to the banks by their roots, their tops angled downstream by the insistent current. Their limbs provide trout a sanctuary from overhead predation—and unfortunately, from carelessly presented flies as well.

Ledges and Trenches

Ledges and trenches are common features in flats, especially in areas where volcanic activity was responsible for much of the geography of the stream bottom. Trenches form holding lies wherever they're a foot or two deeper than the rest of the flat. At times, such lies seem to be crammed with trout, as the entire population of trout that live across the flat might shelter wherever they can find concealment when they're not feeding.

Trenches usually show up on flats as slightly darker areas on the surface. Sometimes a trench is revealed by a slick that is smoother than the rest of the surface of the flat around it.

Reading water in flats depends to a small extent on your ability to recognize all the above types of holding lies. But it depends more on your ability to spot rising trout on feeding lies or visible trout lower in the water column, either feeding or resting.

FISHING STRATEGIES

Fishing flats can be just as frustrating as it sounds like it could be. Trout rise all around you. They give all sorts of clues to what they might be taking. They respond with all sorts of refusals to whatever you are certain will fool them this time. Hope springs eternal. Keep watching. Keep changing flies and tactics. Keep studying. The more you learn about the insects, their imitations, and presentation methods, the more often you'll be able to fool trout feeding on flats.

Observation and change are the elemental secrets to success when fishing flats. Once you learn to observe closely what the trout are taking, and change flies and presentations until you match both the insect and its behavior, you'll begin to unravel the mysteries of flats. One of their most charming mysteries is that they don't get unraveled on every occasion. There will always be times when you, or at least I, cannot solve the prevailing problem and fail to catch the selective trout.

When trout do not feed actively on flats, you can fish for them in the obvious lies: around boulders and logs, in trenches or along ledges, and over plant beds. But the most challenging fishing on flats happens when you spot trout working and fish for them with dressings that imitate the naturals they're taking at the time. On those occasions when everything magically weaves together, you'll find the most rewarding fishing on flats. As time goes on, and you gain experience at this type of fishing, the scarcity of those magic moments diminishes, though I can promise that not every situation on a flat always gets solved.

Tackle

Your tackle for flats must be the finest you can acquire, and your tactics must employ the greatest finesse you can muster. It takes lots of practice to learn to fish flats with more than random success. Even experienced fly fishermen who have achieved their wings on other types of water—riffles, runs, and pools—encounter skunks when they first begin to fish flats.

Reading Hatches

The ability to read broad flats for likely lies can help you position yourself for late-evening hatches such as those of the eastern green drake. When the big duns begin to emerge and make their long drifts downstream, you're in the right position to cast over the trout that rise to them. In this case, Paul Weamer, coauthor of *Pocket-guide to Pennsylvania Hatches* and owner of the TCO Fly Shop in State College, has placed Masako along the stretch of deeper, darker water along the tree-shaded bank of Penns Creek. Trout don't need those shadows in the hours when the green drakes hatch, but it might be the shade that holds the trout there during the day while they, like you, are waiting for the hatch to start.

The big green drake dun can turn a flat from what seems like a blank slate into an eruption of rising trout.

Your tackle should be fine, with the best being $8\frac{1}{2}$- to 9-foot rods in the 3-, 4-, or at most 5-weight class. Floating lines suffice for all fishing on flats; you'll never need to get down so deep that you can't do it in the length of a leader. The line can be either a double-taper or weight-forward, whichever you prefer to use most of the time in other types of fishing. If it's a weight-forward, it should have a long, delicate front taper and a long belly so you can mend with it.

Leaders for fishing flats start at around ten feet long. If you use split shot or putty weight on the leader for nymphing, place it at the end of this and then add a foot to eighteen inches of tippet that balances the size fly you have chosen. If you fish emergers or drys, add three or four feet of finer tippet, again to balance the fly, and you will have a thirteen- to fourteen-foot leader. That's about the right length to start, though you might want it longer. Whatever its length, keep your tippets long, no less than three feet, sometimes four or five feet.

Your tippet will tell on you if it's wrong.

I once watched from the top of a high bank, waiting with my camera, while Jim Schollmeyer fished over a heavy trout rising steadily on a Bighorn River flat. The insects it took were easy to see; they were a size 18 black caddis, which Jim matched with a dark Snowshoe Caddis.

Jim cast quartering downstream to the trout, with a wiggle cast that produced a slack line. His presentations were perfect. The fly fed out ahead of the line and leader and should have appeared over the trout with no seeming attachment. But the fish refused every drift of the fly. Finally Jim stopped casting and started fiddling with his gear.

"Changing flies?" I called down.

"No, changing the tippet from 5X to 6X," he answered.

I waited to see what would happen. On the first cast, the trout rose with great assurance and took the same fly that had passed over it the same way a dozen times before. The trout instantly ran into the plants and broke that fine tippet. It's a common dilemma on flats: You can't hook trout on a tippet that's stout enough to hold them.

Fishing the Bottom

Trout feeding along the bottom on flats are perhaps the most difficult of all. We do not normally think in terms of selectivity when trout feed deep. In faster water, they usually take whatever insect, or reasonable representation of one, tumbles to them on the current.

A couple conditions make that less likely on flats. First, the narrow range of aquatic insect and crustacean species, along with their extreme abundance, makes it common for trout to see nothing but a lot of exactly the same thing. They key in on the natural's shape, size, and color and won't take anything that doesn't look a lot like it. Second, the slow and clear water on a flat gives trout a good look at whatever they're about to eat. If the imitation is not fairly close and is not presented in the same way the naturals arrive to the trout, they will detect the difference and reject the fly.

Fly selection for fishing the bottom should be based on the most abundant food form available beneath the water. This is sometimes easy to check out; just reach down and scoop up a handful of plants. Whatever is most abundant will be crawling around in your hand. But it often happens that what is most abundant is not most available. In such a case, it might become necessary to hold a screen net down along the bottom and wait patiently for a few minutes to see what the current tumbles into it, which will also be what drifts most often into the view of bottom-feeding trout.

You can also accomplish the capture of a sample by fooling a mythical first fish and killing it to examine its stomach contents. Do this only if you desire a trout dinner and are not on water where it's either illegal or would be damaging to the trout population to kill the fish. Some people feel that it's always damaging to the population to kill a trout. It's hard to argue with that logic. It's certainly impossible to argue against the idea that killing one is damaging to that particular trout.

You might prefer to use a throat pump to exhume whatever the trout has eaten lately, thereby springing the insects but sparing the fish. If used correctly, taking a sample this way is not harmful or fatal to the trout, though it does deny it a meal. Don't use a pump on trout less than a foot long. Corral the trout and gently hold it upside down. Fill the pump half full of water, slip the barrel down the trout's gullet, squeeze the tube to inject the water in the bulb, and remove the tube, and it will inhale whatever the trout has been eating and reveal it to you. Release the trout quickly. Be very careful; trout are both fragile and valuable.

What you're most likely to see, if you can get a sample by any means, is something small: a mayfly nymph, caddis larva, midge pupa, or crustacean. You should match it as closely as you can, first in size, then in form, finally in color.

Your nymph should be at least slightly weighted to get it down. If the water is deep and has much current, a split shot or some putty weight might be needed a foot or less above it to keep it from wafting along a foot or two above the bottom, out of the notice of trout. A tiny hard or foam strike indicator, or a bit of yarn placed on the leader, will help you detect takes, although at times it's possible to observe the flash of white as a trout opens its mouth to take your fly.

It would be easiest to fish flats if you could wade into position straight downstream from feeding trout and make your casts from that position. This would reduce any chance they might see you. But you cannot cast from there; the sight of your line and leader flying over their heads would put them to flight. You have to stalk carefully into a position off to one side, from a forty-five-degree angle downstream to almost straight across from the trout.

Cast upstream from the trout far enough to give your nymph a chance to sink down to it. Let the fly dead drift along the bottom or just inches above it. Watch your indicator for any hesitation. Be persistent; it's rare to get a perfect presentation on one cast out of five. Even then, the trout might have just done in a natural and not be quite ready to take again at the instant your fly is ready to be taken.

It requires a lot of observation, and a lot of casts, before you begin to take bottom-feeding trout from flats with anything approaching happy regularity.

Fishing the Mid-depths

If a flat has plant growth rising up from its bed, its mid-depths become an extension of its bottom and can be very productive. Insects or crustaceans live in those plants and may wind up loose in the central layers of the current whenever they lose their grip, entering the drift to redistribute their dense populations, or may get nosed out of sanctuary pastures by marauding trout. Mayfly nymphs, caddis pupae, and midge pupae also are taken at mid-depths on their way toward the surface for emergence.

Whether trout feed on them in the drift or on the rise, it's important to determine just what stage of what minor animal is attracting the attention of the trout. Once you have this figured out, imitate it with a traditional winged wet fly, a wingless wet fly or flymph, a soft-hackled wet fly, or a nymph. A small beadhead imitation of the natural, with just a few turns of nonlead wire to lock the bead in place, will usually sink at the right rate to fish the mid-depths. If you're on a heavily fished flat, it's beneficial to remember that trout have seen lots of bright beads in the last few years; a black bead will give you weight and also a subdued imitation that won't cause educated trout to back away.

You'll almost always fish the mid-depths on flats at the prompting of trout you've seen feeding. You can fish upstream, the same way you would fish the bottom, but leave off the split shot or putty weight and shorten the distance between the strike indicator and fly so the imitation passes through the water at the depth of the feeding trout—or pod of them, as you'll often see more than one feeding on a flat. Take up a position at an angle down-stream or straight across from the trout so your fly is presented to them at the end of the cast, on a straight leader, without anything to alarm them passing over their heads.

You can also present your flies at mid-depths from an angle upstream from the trout. You'll have to wade into position more carefully, because any wading waves will be delivered downstream ahead of you to warn them. Work to where your cast is angled forty-five to sixty degrees down-stream to the fish. Mark a single feeding trout. Cast above and beyond its position, let the fly sink, then coax it into swimming slowly right in front of the nose of the trout. With any luck, you'll feel a slight tightening as the line bellies in the current. Set the hook by raising the rod slowly and gently, or you'll jerk the sunk fly away from the fish.

Fishing the Surface

There are two slightly separate stages to consider when fishing the surface of a flat. One is the moment during emergence when the natural insect reaches the surface film. It will suspend for a few seconds, and trout some-times feed selectively on this stage of the natural during a hatch. You can see the adults on the surface but not the emergers just beneath it. The other is the adult stage after the emerger has penetrated the film and cast its nymphal or pupal cuticle. When trout are concentrating on this stage of a small insect on a flat, true dry-fly fishing is at its most challenging.

If you fail to distinguish the two stages and fish the wrong one, you are in for fishing that is not only challenging but frustrating. Fishing flats requires more observation than any other trout-fishing situation. One way to differentiate the two stages is to follow the drifts of a few adult insects through a pod of rising trout. If none disappear but you see rises right around them, suspect that the trout are taking emergers and ignoring the floating adults. It helps to carry small binoculars to make such observa-tions.

Tackle and tactics for the two kinds of fishing, over emergers or adults, are exactly the same except for the flies you use. Emerger dressings should be extensions of nymphs, with small wing clumps of fur or polypropylene yarn added to represent the half-extruded wings of the natural and also to float the flies flush in the surface film. These flies were covered in my *Handbook of Hatches*, along with imitations for fully formed adults, and you should refer to that book for more detailed notes on them.

When fishing dry flies on flats, your imitations should be the most exact you can either tie or buy. Trout are not likely to accept dressings that are the wrong shape, not near the natural in color, or two hook sizes too large. Nor are they going to be delighted by presentations that are made with tackle that is too coarse to fish as delicately as required on flats.

A position directly downstream and a presentation precisely upstream will accomplish one major event on a glassy flat: It will deliver the fly to the trout you're trying to catch after the line and leader have already passed over its head and astonished it. The best presentation on smooth water requires wading into position either across stream or upstream from the trout.

If your position is across stream, then make your presentation with a *reach cast*. Aim the fly to land in the trout's feeding lane, two to five feet upstream from its lie. While the line unfurls in the air toward the trout, tip the rod over in the upstream direction and reach your arm to its full extent to follow the rod. The line will land on the water cutting at an angle downstream to the trout. As the float of the fly unfolds, follow the fly with your rod and you'll get an extended drag-free drift. The fly will arrive to the trout behind the line and leader. The trout will not be warned.

If you take a position at an angle upstream from the trout, fish downstream to it with a *wiggle cast*, keeping your profile low enough that the trout cannot see you. Aim your delivery stroke to place the fly in the same position: in the trout's feeding lane and two to five feet upstream from its lie. Carry extra line in the air. As the cast unfolds, wobble your rod tip briskly back and forth. The line will land on the water with the S shape of a swimming snake. These curves will straighten out as the current draws the fly downstream toward the trout. The imitation passes over the trout ahead of the line and leader, with no drag. If the trout refuses, let the fly pass, lead it off to the side, then pick it up and make another presentation.

It usually takes many attempts before everything becomes perfect and the trout takes the fly. Count a beat before setting the hook or you'll pull it away from a trout you would very much enjoy catching, since it's among fly fishing's most difficult.

9

Probing Pocket Water

Any object that interrupts the current can form a quiet bit of water where a trout can hold a station, with a window on a territory that feeds it well, even in otherwise violent flows. Most prime lies in riffles and runs are pieces of pocket water, in a way. But in this chapter, I want to treat the kinds of prime lies that are more commonly considered when we use the term *pocket water*.

Pocket water is any quiet hesitation that holds trout in the midst of water that is too fast and turbulent for them to remain anywhere else for more than a brief dash to intercept something from the drift. Riffles and runs have pockets of quiet water, but the water around them is usually not quite so brutal that trout can't hold in it when given a good motive, such as a lucrative hatch of insects.

As defined here, pocket water is found in rapids and cascades. The current in these water types is so swift that trout cannot survive long without something to hide behind. A trout tires of constant fast swimming in less than five minutes. A trout fighting frantic water would have to swim urgently upstream to keep from being delivered downstream like an aquatic insect awash on the drift. Trout also do not hold in water that is constantly seething, which would cause them to be unable to orient themselves and hold a steady station. Pocket water, then, is a comfortable lie in surrounding flows in which a trout would not be able to find and hold a station.

STRUCTURE

Pocket water is found only where the streambed has a steep gradient. Steep water is not restricted to mountainsides. Almost all moving waters have stretches that bolt through gorges, rush down through tumbles of rock, or cut drops through ancient lava flows, with lots of rapids and cascades. Most trout streams offer some reaches of pocket water fishing, and it will

When you're wading in the midst of pocket water, you're essentially looking for dark, relatively flat water, rather than white, violent water. A productive pocket can be anything from the size of a basketball on the surface, indicating a place where the water pauses after it's gone around a boulder on the bottom, to an area as large as the one I am fishing here on the North Santiam River in Oregon. Not all pockets that hold trout reveal themselves on the surface. Many boulders along the bottom form holding and even prime lies without sending a soothing flat spot to the top.

If you're fishing pocket water more wadable than this, you might find it worthwhile to move slowly and carefully, fishing all the water that looks as though it has potential to hold trout. In that way, your flies—usually a bushy dry or heavily weighted large nymph trailed by a smaller beadhead or Copper John—become the tool with which you read the water. It's often overlooked, on the water and, I'll confess, in this book, that your flies become the eyes with which you read trout water.

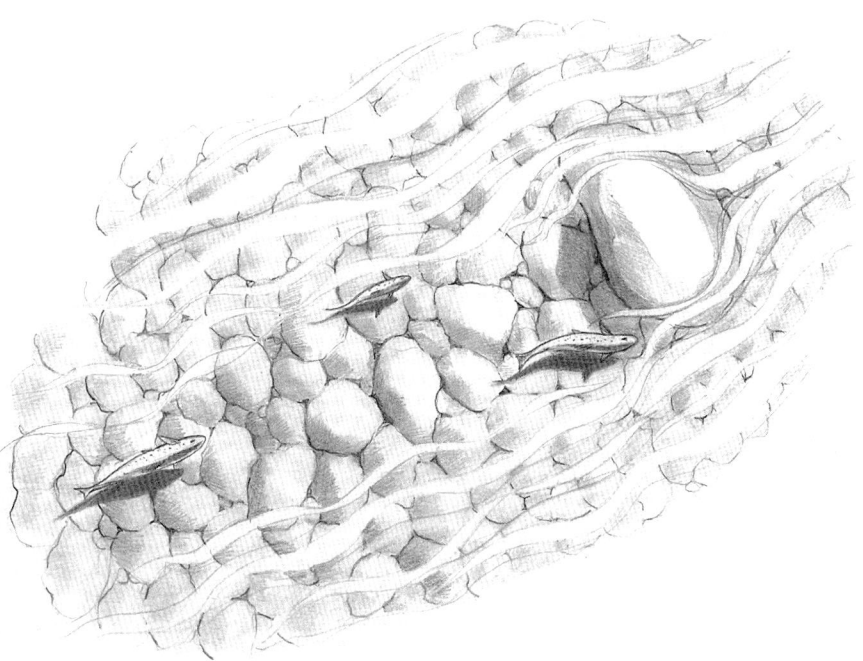

usually be good fishing, because pocket water is so difficult to fish that most anglers avoid it.

We tend to think of boulders when we think of quiet spots in rushing waters. Boulders are the most common cause of pockets that hold trout, simply because they're about the only thing heavy enough to hold the bottom in strong currents. A log or limb rarely gets a chance to settle down in such fast water, though if one gets caught between boulders, it then forms pocket water that is excellent.

Boulders of sufficient size to form holding lies in rapids and cascades usually, though not always, protrude from the water. They sit out there calmly with water frothing all around them, stained white by bird poop on their tops. Water piles against a boulder's upstream side, tries to shove it downstream, gives up and swings around it, leaving an eddy in its wake. The eddy is a swirl on the surface, but there is at least one quiet spot down beneath it, usually right on the bottom, where a trout can rest and wait for insects and other forms of food to tumble by.

Ledges and trenches also form in fast water. Some slice across the current, marked by a line of whitewater followed by a patch of quiet water tucked up against it downstream. Others run parallel to the current, marked by brief slicks in the fast water where the surface reflects a bottom that has dropped away.

Almost all pocket water lies are formed in one way or another by boulders. Some protrude and are easily seen, and the pockets they form are easily read, like this one on a plunging mountain stream in the Andes of Chile. Many boulders do not protrude and send either slight boils or small patches of slick water to the top.

An occasional abrupt leveling in the steep gradient of a set of rapids will also slow the water, creating a very brief pocket run in the midst of water that tumbles brutally upstream, downstream, and off to both sides of it. These look remarkably like miniature runs, with level tops and unbroken flows for ten to perhaps thirty feet. Because the water all around them is rough, they are normally quiet spots in the storm, but not as smooth on top as they would be if they were surrounded by quiet water. You have to look carefully to separate these short pieces of holding water from the uninhabitable water that is all around them.

The size of a piece of pocket water depends first on the size of the stream, second on the size of the structure that forms the pocket in the stream. A giant boulder in a large river might make a pocket twice the size of the largest plunge pool in a mountain creek. A piece of pocket water in a mountain creek, in contrast, might hide behind a boulder the size of a basketball and give you a casting target no bigger than a basketball hoop.

NEEDS OF TROUT

The needs of trout are met in pockets in largely the same ways they are met in the prime lies of riffles and runs.

Shelter from Currents

Shelter has been declared by biologists to be the limiting factor to the value of a holding lie in all types of water. That is especially true in pocket water. Trout hold only where they find water calm enough to give them some comfort.

The obstruction that creates a pocket gives trout shelter from the swift current on the downstream side. The upwelling of water on the upstream side of the same obstruction is not often slow enough to form a pillow of calm water. The best holding water in a pocket is almost always found downstream from an obstruction, deep along the bottom.

A ledge or shelf large enough to be reflected to the surface as a bit of smoother water amid the storm will also have enough quiet water down below to shelter a trout or two. Sometimes a pod of trout will hold in a fairly large trench, ten to twenty feet long and two to three feet deeper than the water around it. When you get into a patch of this kind of water, you can catch half a dozen trout on the same number of casts, all the time wondering why you're suddenly such a superb fisherman. It's because you read the water right.

Protection from Predators

Protection from predators is excellent in pocket water. The surface is too broken for vision from above; the water is too broken for predation from below. Angling predation is discouraged by the kind of wading needed to fish pocket water right. If trout find shelter from currents in rapids or a cascade, they'll usually also be safe from most predators.

Trout Foods

If a pocket is large enough to provide shelter for a trout, it's almost sure to be large enough to provide food for it as well. The fast and oxygen-rich water of rapids or cascades can be kind to insects. The rocky and bouldered bottom has lots of crevices and spaces where nymphs and larvae find protection from the current. But because the water is swift, lots of these same insects get knocked loose and delivered in the drift.

A slick on the surface, such as this one on Montana's Boulder River, outlines a trench where the bottom has fallen away. This one is large; some are very small. Learn to watch the water carefully for them, as they always mark prime lies. The lip of the trench breaks the current, forming a soft spot; the deeper water makes it impossible for ospreys to dive in there; and the constant current passing overhead drops plentiful food into the trench. Sometimes you'll find what seems to be a pod of trout in a trench of any length. Because of the laws of territoriality, which increases as water speeds up, it's more likely that the trout are lined up the length of the trench than that they're podded up in one place. But it's not uncommon to catch several trout in a single trench, especially when it's surrounded, as this one is, by whitewater.

During the warmer months of the year, when trout and trout fishermen are both most active, insects that have reached maturity constantly migrate out of fast water in preparation for emergence where the water is more congenial. They arrive in pockets for the same reason that trout hold there: for shelter from the fast water surrounding the pocket. They need a quiet place to perform their emergence. These migrating insects are safer from the current when they emerge in pocket water. They are not nearly as safe from trout.

Trout foods in pocket water are primarily aquatic insects. Mayfly nymphs in rapids and cascades belong to the clinging variety, with some crawlers holding on as best they can. The predominant species fall into the medium range of sizes, 12 to 14, and some are even larger.

Most caddisfly larvae that live in fast water construct their cases of sand and pebbles. The ballast of the heavy material holds them on the bottom, which is where they want to be. They're generally medium to large insects, size 8 to 12.

Stonefly nymphs that live in fast water are generally large salmon fly and golden stone varieties, adapted well to crawling about among the stones but poor swimmers when dislodged by brisk currents. These are the primary reason heavily weighted size 6 and 8 nymphs work well when pitched right to the bottom of pocket water. But they're always most effective when trailed by smaller nymphs, size 14 or 16.

Many aquatic insects move into the relatively calm water of pockets when it's time to emerge. More of them migrate through such pockets on their way to edge currents, where the water is even softer. Trout see fairly heavy numbers of a single species at rare times, but it's not usual for pocket water trout to be selective. Fast water, and the fractious way it delivers drift, gives a trout little time to light up its pipe and think things over when a promising bit of drift tumbles its way. The trout must dash for it, grab it, and make a decision about it—is it an insect or a bit of bark?—after its flight downstream has been contained. You have to be alert, and set the hook quickly, when fishing pocket water. Trout are used to stopping things to test their edibility and spitting them out if they're not. Your nymphs are not edible, no matter how prettily you tie them.

If a stream is narrow, quite a few terrestrial insects find their unhappy way into the drift. If the water is swift and rough, few of them float. Most are taken by trout beneath the water, as part of the drift, after they've drowned. When terrestrials are most abundant, though, trout usually focus on the surface and cock their fins, ready to race up and spear a beetle, termite, or inchworm before it gets swept away or taken by a faster trout. Then you can catch pocket water trout on dry flies.

Temperature and Oxygen

Temperature and oxygen regimes are excellent in broken pocket water. If trout are forced by a summer heat wave to pack their bags and move out of one water type to find comfort in another, pocket water is where they go. Under these conditions, especially if poorly oxygenated pools are located nearby, you sometimes find pocket water holding a richness of trout.

TROUT LIES

Any pocket that holds trout in fast water forms a prime lie, meeting all the needs of the trout in that one place. The size of the trout will be directly related to the size of the pocket and the amount of food that it provides. The bigger the pocket or the richer the stream, the bigger the fish.

Reading pocket water is a two-pronged affair of looking for obstructions in the current and learning to spot slight slicks and eddies that indicate still pockets beneath the current, formed by ledges or trenches. One easy way to think about the fishability of a pocket you're scoping is to envision a bushy dry fly cast to its surface. If the water is gentle enough that the fly would get a good float, even if only for a foot or two, the pocket is large enough that it might hold a trout. If the water descends so harshly onto an obstruction that it tears whitely around it and no fly would ever float behind it, then it's unlikely that a trout could hold there either, though one might find a soft spot right on the bottom.

FISHING STRATEGIES

Trout are always on their stations in pocket water, hanging tight along the bottom, in the water just downstream from whatever obstructs the current. They are nearly always prepared to feed. The distance they'll move to take a fly depends on the depth of the water, its vigor, and any insect activity that has rewarded them recently so that they are inclined to move and feed.

Safety is always a concern when fishing pocket water. Carry a wading staff to help you keep your footing. I like a folding staff that fits in a belt holster. It's out of my way until I'm teetering and need it. Then I draw it like a sword and stab the bottom at the last instant to save myself from tipping over. I haven't been nominated for knighthood yet, but neither have I gone swimming in rapids or a cascade, in which water types I usually keep the staff drawn and dangling in the current downstream from me, except when I move from place to place and use it. I don't like rigid staffs; when they're not employed saving me, they seem to keep busy trying to trip me.

Reading prime pocket lies is a simple matter of looking for obstructions to the current, then examining the water for places that look as though a trout could hold along the bottom there without dealing with a lot of turbulence. Trout do not like to be tossed around and will not take up holding lies in seething currents, though they will move into them to feed. In this photo of the Chilean stream rushing off the Andes, it's easy to read the prime lies to the left of the white current tongue in the foreground and in the slick on the bench above, the one I'm probing with my nymphs.

You might wonder why my examples of pocket water come from such distant streams. I wondered myself until, sorting through my slides, I realized that one of the unwritten rules of pocket water fishing is not to carry a camera for fear you'll dunk it, so I rarely carry one when fishing brutal streams unless I'm in some exotic location.

Wading safety should be your first concern when fishing pocket water. Trout are not quite as important as we make them out to be, though I'll admit they're high on my list, or I wouldn't be writing this book about how to find and catch them. But I just lost a friend, Dr. Keith Hansen, who was swept into a logjam while wading steelhead water on the Dean River in British Columbia.

Whenever you're wading pocket water, wear stout wading brogues, cinch a wading belt around your waist, and always have a strong staff on a cord exactly at arm's length. Also, as an extension of my rule against carrying a camera, I recommend that you not carry anything extra. If you can get into a position to brace yourself against a boulder or use one to stabilize yourself in all that unstable water, as I have in this photo, then do it. Make sure your feet are solidly placed before you turn your attention to casting when the water around you is violent, as it is in much of the best pocket water.

It's best to avoid getting tripped in whitewater. It's never fun and can be fatal. Wear chest waders with a wading belt tugged tight and felt-soled wading brogues that are solid enough to protect your feet from constant pounding. If the bottom is slick, a sign that it's rich in feed for insects, wear some sort of studs that cut through the algae. Wade slowly, always planting one foot and your staff before moving the other foot.

You take a chance every time you fish this kind of water. Never increase the risk by wading where you shouldn't. Lots of pocket water is so formidable that it's simply sanctuary water for trout. Leave it at that.

Tackle

Tackle chosen to fish pocket water should be what you would carry to fish the rest of the same stream. Pockets are opportunity water; few streams have pocket water as the dominant form of holding lies. You want to be prepared to fish pockets when you come to them, but you don't necessarily want to go to a stream just to pick its pockets, unless it has lots of neglected pocket water that holds lots of neglected trout. Then it's worth it to arm yourself specifically for the pocket water.

If you're prepared to fish the riffles and runs of the creek, stream, or river you're on, you'll also be prepared to fish its pockets. Everything tends to be relative. But pockets and pools have one thing in common: If you're lightly armed for the rest of the water, it will show up as a handicap when you approach a pool or piece of pocket water.

If a rod were chosen just to probe pockets, it should be long and stout enough to be bossy. The line should be a 6-weight to command heavy flies. There is no difference between double-taper and weight-forward lines, because all your casts will be short. The line should be a floater. Leaders should be no longer than the rod and stout. Tippets should always be heavy enough to turn over fairly large flies and strong enough to hold trout against strong currents.

Fishing the Bottom

The bottom is the most likely level to probe first in pocket water, unless the water is shallow and insects are active, in which case a dry fly might be most effective. Dry-fly fishing also might be the most fun, and there is nothing wrong with pecking away at pocket water with a dry even when it's not the most effective method. I do it all the time. But I catch far more pocket water trout when I switch to nymphs.

It's easy to notice that in every photo of fishing pocket water, the rod is held high, as in this photo of Masako on Utah's beautiful Logan River. The reason is simple: to carry the line and even a bit of the leader, if possible, off the water, so the swift and always conflicting currents can't grab them and whisk your fly away from the water you're trying to fish. The closer you can wade to the water you want to fish, the fewer of those currents you'll have to solve. The higher you hold your rod, the less line you'll have on the water for currents to grab.

Wading close and high-sticking are very important in pocket water fishing, considering what I call the "worry distance" of the trout: the distance at which their senses warn them that you're near and you're dangerous. That distance is cut to one or two rod lengths in rough water.

When fishing the bottom, use weighted nymphs. Select the size based on your estimation of the typical trout foods you think might drift into those pockets. On small streams, use nymphs in the size 12 to 16 range. On medium and large streams, where you would expect pocket water trout to go for larger bites, use nymphs tied on size 6 and 8 hooks, but drop a smaller one off the stern. The larger dressings should be salmon fly or golden stone imitations, the smaller ones standard searching patterns, most of which will benefit from the addition of tungsten beads for a faster sink rate.

Rig to fish nymphs with a large, hard strike indicator; pockets are no place for yarn or small foam indicators. Balloon indicators are excellent. They float high and are very visible, but they're not obtrusive, not that it would matter in a pocket. If the water is clear enough and you'll be fishing short enough casts to see your indicator while it's submerged, fix two small ones to the leader a foot or less apart, a bit less than the depth of the water up the leader. Use nymphs with lots of weight. Watch the indicators: If one suddenly moves in opposition to the other, set the hook. That's a trout.

Present your nymphs near the bottom and as close behind the obstruction that forms the pocket as you can possibly get them. Getting them to plunge swiftly might require some extra weight pinched or puttied to the leader, eight to ten inches above the top nymph or even between them, if you're fishing two. There is no substitute for fishing as close as you can to the pocket, which allows you to keep in close touch with the drift of your nymphs.

Wading is difficult in and around pocket water; use your wading staff. If you can reach a position from which you are able to loft your rod almost over the pocket, you'll be able to lift most of your line off the water. The fly will sink faster if it's impeded only by the leader and not by the line. And you'll be able to direct the drift of the fly by steering it with the rod. This increases your chance of getting a strike and also enhances the chance you will learn about it when a strike happens. If you keep direct contact between the rod tip and your nymphs, a hit to a nymph fished through a pocket might even be telegraphed to your hand by a sharp rap.

This kind of nymphing is referred to as *high-sticking*. Because pocket water is turbulent, you can wade into a close position without scaring trout. Don't wade into the pocket itself. It's better to approach from downstream or off to the side than it is to wade in from upstream. Dirt and debris kicked up by your feet might put fish off their feed. But some pocket water is so strong that you cannot wade against it. If that is the case, let it push you downstream, but don't let it topple you over.

Cast to all the water that looks peaceful enough to hold a trout. In small pockets, this might require only half a dozen casts. In large pockets, you

Reading pocket water from a boat becomes a hectic matter. The water is moving fast. The boat is moving fast. If you move a trout, it will come fast. You have to mind the drift or float of your present cast while at the same time calculating the placement of your next cast. Your vision and attention must be split between where you're fishing now and the best bit of pocket water to be fishing next. Sometimes your guide will be yelling instructions at you, though this day on the Colorado River, we had the luck to be under the calming influence of head guide Andrew Petersen out of Trapper Rudd's Cutthroat Angler shop in Silverthorne.

One thing I've learned from the modest amount of guided boat fishing I've been able to do: You almost always catch trout when your fly is in the water, even if it's in the second-best place it could be, and you almost never catch trout carrying your fly in the air, trying to make sure it hits precisely the best piece of pocket water. Still, it's helpful if you can fish out the drift that's already working for you while looking ahead to calculate the next pocket that's most likely to send a trout out to take a whack at your fly.

should fish the water with a casting pattern that covers all likely places a trout might hold. Fish patiently enough to show the fly to any trout that might be crouched down there. You had to wade hard to get into position to fish the pocket; don't take a couple casts and then burn up energy forging on to the next pocket.

Fishing the Mid-depths

Fish the mid-depths of pocket water only if some sort of insect activity indicates that it would be effective to do so. A hatch of mayflies or a dance of caddis over the pockets might indicate that trout are visited by lots of drift down below. In this case, use a wet fly or small nymph fished on a floating line. Lengthen your casts a little; it helps to be upstream a bit so you can fish downstream at a slight angle to the pocket.

Cast just above and beyond the obstruction that forms the pocket. Let the sunk fly, or pair of them, drift down to the pocket. If it's formed by a boulder sticking up above the surface, mend your line over it, then coax the wets around as close as you can behind it. Let the fly or flies swirl in the eddy. Hold your rod high to keep line off the water so it won't tug the fly out of holding water. Watch the water closely; takes in this kind of fishing are often visible, with a flash of flank as the trout turns or a splash if it takes near or even on the surface.

You can fish wet flies effectively upstream, working your way from pocket to pocket, casting short and watching your line tip for any hesitation as the fly drifts back downstream toward you. It's a little like dry-fly fishing, but it's a lot more demanding. It works well where pockets are no more than two to three feet deep, at times when trout are unwilling to come all the way to the surface for dry flies. I've found upstream wet-fly fishing most effective when the sun is out and bright on the water and trout are for some reason bashful about the top.

Fishing the Surface

I've always written that we fish not just for trout, but for *surprise*. Dry flies bobbing on pocket water provide many of angling's ultimate moments. Drys are also the most effective way to take trout from many pockets. They work best where the water is shallow, two to three feet deep, and when adult insects are active, getting onto the water.

Dry flies for pocket water should be chosen for a high ride and bright visibility. Royal Wulffs and Humpies in size 10 to 14 are traditional favorites. Elk Hair Caddis work well in the same sizes. Stimulators, which imitate golden stoneflies, draw up lots of trout when used in size 6 to 10. All of these flies float well and are very visible to both the angler from above and the trout from below. If you have favorite searching dry flies, never neglect them.

Finesse is not the first requirement in pocket water dry-fly fishing. Getting trout to notice your fly is more important. Smacking the fly to the water will sometimes announce its arrival to the trout, though never do that on the first cast to any bit of water; you might frighten trout you would rather attract. It helps if you can hold as much line off the water as possible. Wade close, cast short, and keep your fly tap-dancing on the eddied water behind a rock as long as you can get it to stay there.

10

Trout Lies along the Banks

Banks are the transition zone between the aquatic and terrestrial environments. The narrow bit of water that abuts against a bank gets enriched from both directions. Many mayfly and caddis species, and all stoneflies, migrate across the streambed and emerge in the calm water very near the shore or crawl out to emerge on protruding boulders or the streambank itself. The adults of these aquatic insects, and many more that emerge out in open water, hang around in bankside grasses and brush and get onto the water often. Terrestrial insect populations live in the same streamside shrubs and grasses. They fall to the water much more often than they prefer. Trout hold tight along the edges, eager to feed on all this focused abundance of both aquatic and terrestrial insects.

Good bank water is revealed by a combination of factors that meet the three basic needs of trout: enough depth to provide protection from predators, at least a fairly strong current to deliver food to the fish, and obstructions along the bottom or indentations in the bank to shelter them from that same current. A good bank has all three of these essential factors. If any one of them is missing, it won't be bank water at all.

Once you learn to recognize banks where these three factors come together, it doesn't take long to learn to read productive water and, just as important, eliminate bank water that will not hold trout.

STRUCTURE

All creeks, streams, and rivers have the full complement of banks on both sides, but not all bank water makes good trout water. To attract and hold trout, a bank must have depth, current, and obstructions to that current. Depth comes first, is most rare, and might be considered the limiting factor. Look for water that drops off quickly, rather than sloping off gently. If it's a foot or more deep on the bank and two feet or more deep a foot or

two out from the bank, it will hold trout. If it's an inch or two deep at the bank, a foot deep three or four feet out, and doesn't reach a couple feet deep until ten feet out, it might be good water out where it's deeper, but it won't be good water along the bank.

The more stable the flow of a stream, the more steeply it will erode its banks and the more trout water it will offer at the edges. The more a stream is subject to high runoff and low summer flows, the more it will push its banks back, leaving wide and shallow gravel bars, and the less fishable bank water you'll find. But when you learn to recognize it, good bank water can be found on stretches of all types of streams.

Meadow streams, especially those with steady spring sources, tend to work their deepest water up against the banks, often as undercuts. It takes water with depth to make good bank fishing; spring creeks typically offer an abundance of it. Tailwaters with relatively stable flows, usually those with reregulating dams downstream from power-generating dams, have many reaches with good banks. Tailwaters that are subject to wide fluctuations in flows, often daily, do not offer the stability to provide much good bank fishing.

Freestone streams, subject to spring spates and summer low flows, offer fine bank fishing in many bends, where the strongest current pushes up against the outside edge, eroding depth there. The inside of the bend will almost always be a shallow gravel bar, the worst kind of bank water

To be good bank water, it's necessary to have current to deliver food, something to break that current, and sufficient depth for protection from overhead predation. This stretch of the Madison River, in Yellowstone Park, offers all of those things. The water is more than knee-deep, the current is somewhat brisk, and both the submerged boulder and that log lodged in the bank serve to shelter trout from the flow.

It's fairly rare to be able to fish bank water by wading out and casting back in toward it. Most of the time, it's fished either by creeping along the bank and casting short upstream or by inserting yourself into the water right against the bank and then wading and fishing very slowly and carefully upstream.

except when predaceous trout are on the hunt, usually at dawn, dusk, or through the night. The average bank in a freestone stream falls short of being good bank water. It's critical to learn to read banks right, because if you were to fish them all, you'd spend much more than half your time casting over water that failed to give you half a chance to catch a trout.

Good banks on meadow streams are easily recognized as excellent water. They're often fished hard. Good bank water on freestone streams is more difficult to separate from bad bank water. Some excellent banks are rarely pestered. At times you'll discover good reasons for this lack of attention; it can be murderous to make your way along a brushy, bouldered bank to tap the best bank water on a typical freestone stream. But such a bank can reward you well.

NEEDS OF TROUT

The needs of trout are met generously along many types of streambanks. They are not met at all along a lot of others. An angler who does not know the needs of trout, and cannot recognize how moving water might fulfill those needs, can waste a lot of time fishing water empty of trout. An angler who can read bank water can make a lot of casts in a day, over very productive water, without ever getting his feet wet.

Shelter from Currents

The need for shelter from the same current that delivers food is met either by boulders along the bottom, obstructing the flow, or by indentations in the bank that let trout tuck in, out of the flow. Banks that provide these sorts of cover usually have either rocky, bouldered banks or a good bit of vegetation. Grassy banks with undercuts beneath them and bites taken out of them offer excellent bank fishing. Bunchgrass clumps often sweep out over deep bank water with a fair to fast current, forming good lies. Willow patches sometimes spring from a single ball of roots. The root ball breaks the current; the overhanging limbs drop shade and a constant supply of awkward insects. Trout find comfort in such places.

Rocks and boulders sloughed off the bank break the current and provide perfect lies. Banks with boulders are usually steep. The water falls away quickly to three to six feet, sometimes right into a fast current. You don't want to wade such water, and it's often better to stay out of it anyway. Much of the best bank water on any stream or river can be fished by boulder hopping up the shore, sometimes thrashing through the brush. I used to fish the banks of my home Deschutes River in hiking boots, though never shorts; I'm far more afraid of poison oak than I am of rattlesnakes, but those Deschutes banks have plenty of both.

Boulders that break the current immediately adjacent to the edge hold trout on both the upstream and downstream sides. Be sure to fish both.

Protection from Predators

The need for protection from predators is met by depth, at least a foot and a half of it, preferably from two to four feet of water within a foot or two of the bank. The same need can occasionally be met in shallower water by overhanging vegetation or a patch of shade that falls directly on the water, but depth is the primary factor in protecting trout from kingfishers when they're small, ospreys and herons when they grow larger. Lack of depth is the primary sign of empty water. Where the bank slopes away gradually, only inches deep, it seldom holds trout, and you should waste little time fishing it.

Trout Foods

The need for food is met along banks from both directions, by aquatic and terrestrial insects. But it takes a current to entrap those luckless insects, compress their numbers in a line along the bank, and deliver them downstream to likely lies for trout, which don't hold everywhere along any bank. If a bank lacks a conveying current, it will also lack trout. The current does not have to be strong, but it has to be at least modest. You need to read current speed carefully; if the water blasts along the bank with too much force, trout will not be able to find any lies where they can get out of it, and they won't be there.

Trout foods along banks run almost the entire spectrum of aquatic and terrestrial insects. Late-spring and early-summer migrations of the largest stonefly nymphs, salmon flies and golden stones, draw trout to the banks. The naturals queue up right at the edges, waiting their moment to ascend the banks and take up aerial life. Trout line up in good bank water and focus their feeding downward on such big bites, and it's often impossible to bring them up to dry flies. A few stonefly adults in streamside vegetation promises trout lined up right at the bank feeding on nymphs and nothing else. Learning to notice such things is one way to learn to read water.

Adult aquatic insects along banks tend to be stoneflies in spring, caddisflies in summer and fall. Mayflies, with a few exceptions, spend little time along the banks, but big swimmer nymphs live in spring creek undercuts and crawl out on grass stems to emerge. If a strong wind blows them to the water, these emerging duns can become significant right at the edges.

Key terrestrial insects along banks include ants, beetles, and grasshoppers. Though these are most important, and can cause selective feeding if a single type is both abundant and available, it's the smorgasbord typical of terrestrials that makes it possible to fish most bank water effectively with searching dry flies. In my own bank fishing, my favorite combination is a size 12 or 14 Deer Hair Caddis with a size 16 or 18 Beadhead Fox Squirrel Nymph dangled off the stern on two feet of tippet.

If trout seem selective but I can't figure out exactly what they're taking, I'll try a size 14 Foam Beetle or size 18 Black Ant. If hoppers fly out from beneath my feet as I approach any stream, I'll use a size 10 or 12 Parachute Hopper. If caddis or stoneflies are dominant, I'll select an appropriate imitation.

Trout Foods Can Predict Trout Lies

Trout lies can sometimes be read by the presence of certain foods. One of my favorite streams, Oregon's upper Malheur River where it emerges from a national forest, has a heavy population of gray and very large grasshoppers in its nearby grasses. They are exactly the color of the rocks on which they often rest. When they take wing, they frequently make mistakes about where to land. Trout move right up to the banks to take advantage of them. By assessing the presence and availability of them, I can decide whether trout will be tucked into bank lies or lurking in safer lies farther out in the stream.

It's interesting that this changes entirely the way this stream must be read. The first time I fished it, one early September day, I naturally focused on all the easily read central lies, deep runs with current tongues promising food delivered at their upstream ends. But these lies appeared to be essentially empty of trout. At lunch, I was still relatively skunked and wondering why, on a stream that not only looked as though it would hold lots of trout, but also was rumored to have provided them in abundance to others who had fished there. While I ate, I noticed that those pesky gray hoppers were beginning to get warm in the sunshine, taking some experimental flights. They'd land on my wadered legs or plunk down in the middle of my meal. That gave me the idea

to fish bank water, though it was far shallower and looked more exposed to danger than those deeper central currents.

It didn't take long to realize that trout had moved to the banks without leaving many friends behind. All were there, lined up in any water that was deep enough and had current enough to be even marginally safe. And they were quite willing to intercept big, dead-drifted hopper patterns. It surprised me how gently they would rise and pluck them under, so subtly that I often noticed no more than the absence of the fly I'd been following on its drift. Some of those trout responded violently when I raised the rod to see if they were out there. A few pushed three pounds, large for such small water.

Temperature and Oxygen

High water temperatures and consequent low oxygen levels are rarely limiting factors in trout deployment along the banks, except in cases where the problem is already pronounced on the rest of the stream or river. If the entire system is too warm, then water along open banks will be too warm as well, and you won't find trout there. They'll be withdrawn to cooler depths, if there are any, or toward the lower ends of riffles, where the water is well charged with oxygen. They'll also migrate toward any springs that upwell from the bottom, whether out in midcurrents or along the bank itself.

Shade cast along banks would seem to create holding lies by cooling the water. That would be true if the water were still, but it's not. Bank water that holds trout is defined by sufficient current to deliver food. If it's fast enough to escort food along, it's also fast enough to move in and out of shade and sun quickly enough that it's neither warmed significantly where sunstruck, nor cooled sufficiently where shaded, to affect temperatures in a specific holding lie. If a long stretch of bank is shaded, however, that entire bank might be cooled enough to draw trout to it. Certainly, if all else is equal and temperatures aren't high enough to drive trout to seek cooler or more oxygenated flows elsewhere, they'll be found in shade along a bank as opposed to an exposed area whenever the sun is overhead.

The one temperature-related circumstance that can cause trout to gather along a bank dependably enough to be plugged into your practice in reading trout water is a cooling tributary that enters a warmed stream or river. If the edge water downstream from the confluence is shaped correctly to hold trout—again, depth for protection from predators, current to deliver food, and obstructions to the current that delivers that food—then trout will move from other parts of the warmed stream or river to hold in that one stretch of cool bank water. Such situations will not be common. If you find one, you might find it a place in which natural territoriality has been suspended, and an outsize number of trout can sometimes be taken from a short stretch of stream.

As always, when water temperatures are high enough, and oxygen levels low enough, to cause trout to be stressed without being tugged around on the end of a string, leave them alone, even if you find them temptingly easy to hook.

TROUT LIES

You'll see several tip-offs to the kind of bank water that holds trout worth struggling to get to. An undercut bank is an obvious one, especially on an outside bend that has sides falling steeply to the water. Hummocks of

It's always wise to remember that bank lies are right at the bank, six inches to no more than two feet out. That is good bank water all down the length of the half-submerged log on the Metolius River, and the rock tucked against the log forms prime lies both above and below. The boulder in the currents a few feet out from the log is also prime; it just isn't defined as a bank lie.

bunchgrass or clumps of willows increase the chances that this will be productive water, because their root systems grip the soil and form overhangs and indentations. Gentle boils working along the bank indicate rocks just under the water and increase the number of lies for trout. When you fish such trembling banks, place your footfalls very gently or you'll frighten the trout before you get into casting range.

Man-made riprap banks are almost always productive. They are in place to stabilize the banks and are seldom wasted on water where the currents are not brisk, causing the risk of erosion. Riprap is best wherever the rock was dumped casually, allowed to roll down the bank and settle as it might. The resulting underwater jumble has lots of prime lies. Riprap is not so fine where it is prettily laid, like a paving-stone roadway tipped on its side, half in and half out of the water. Such banks provide few holding lies.

Natural rock banks, caused by tumbles of boulders off the hills above the water, are prime areas wherever they have a fair current sweeping along them. But natural rockfalls usually are wasted if they step off into still water. If the water in such a case is more than five or six feet deep, however, it might make a very fine trout pool, and the tumbled boulders will increase that chance. Such still water should be fished with tactics described for pools, not for banks.

Riffles rarely have productive banks unless they flow through narrow slots and hold their depth right against the shore. Most riffles have shallow, sloped banks, and their trout hold farther out. Runs often have productive banks, with the water a couple feet deep and flowing well right against the shore. Wherever such water has sufficient indentations or breaks to the current under the surface, you'll find trout.

Pools can be good bank water if they're deep right to the edges and have some current there. But it's usually most productive to fish pools from the inside of the curve, wading in shallow water and casting long toward the deeper outside bank. Normal pool tactics work well in these situations, and bank tactics are not as often the best bet. Bank tactics should be employed only where the pool is so wide that you can't wade the shallow side and cast far enough to reach the deep side. Then you can work your way upstream on the outside edge and fish it as bank water.

Flats generally have excellent bank water, though it's usually overlooked. Where a flat keeps its even depth to the very edge, the banks will often be the most productive water because of all the food that gathers there. Anglers tend to concentrate on hatches that happen in the open water of a flat, fussing over the mayflies and midges that emerge there. That's the way it should be; flats are supposed to be frustrating. But turn your back on those rising trout long enough to examine the transition line, and you're

Some bank lies, formed by boulders along the bank, are very comparable to pocket water. Trout are almost compelled to take up residence in them by the forcefulness of the currents outside of them, in the main stream. This bank pocket, from which Masako has picked a trout, is on the Boulder River in Montana.

likely to find trout rising there too. Sometimes they're larger than the ones rising out toward the middle. Often they feed on a potpourri of aquatic and terrestrial insects and are less selective than the others. Just as often they focus on some terrestrial minutiae, and then they might drive you back out into open water, where things seem simpler. But a tiny beetle or ant dressing will almost always fool them.

You might not consider pocket water to be a feature of banks. Yet you'll often notice pockets in such swift water that you dare not wade it, but you can reach some of the pockets with a cast from the bank. There are prime lies on many tumbling reaches of river where the biggest trout are picked out from behind boulders near shore, while the water away from shore, out of wading and casting range, is so brutal its trout are absolutely safe.

FISHING STRATEGIES

Trout, when they're found along the banks, are like Napoleon: They have not come to make a speech, Josephine. They are there to feed. They're accustomed to making quick decisions. Because most of the feed along the transition line arrives on the surface, they are usually eager to ambush dry flies. Because their supply lines seldom consistently provide identical groceries from one moment to the next, they are not usually selective, though they can be.

Successful tactics for fishing banks call first for working your way into a good position from which to present your fly. This means you might have to fight briers and brush and, in some places, such as my home Deschutes, even worse. Rick Hafele fished the Deschutes with me last spring. He came back to camp for lunch one day with eyes that were rather round. I asked him what had happened.

"I got going down a steep gravel bank to the river too fast," he told me, waving his arms. "I started skidding. I heard something buzzing and looked down. A rattlesnake and I were on a collision course. I managed to backpedal and tread gravel. The snake must have been as panicked as I was and hurried for its hole. It dove underground and disappeared." If that snake were as wide-eyed as Rick when it got back to its camp, it was probably still waving its own arms, telling its friends about its narrow escape.

I often wade banks, slipping in and out of the water, moving upstream, stabilizing myself with handgrips on streamside vegetation. I learned to cast left-handed so I can fish both banks of the Deschutes River with an equal chance to catch trout and clutch handholds to avoid being swept away. It's a brutal river.

Tackle

Most bank fishing is dry-fly fishing, and your tackle should be chosen with that in mind. Your gear should be fairly standard for fishing dry flies anywhere. It's rarely presentation fishing, with fine tippets and tiny flies. You don't want a soft rod. I use an $8\frac{1}{2}$-foot rod with a fairly fast action, balanced to a 5-weight floating line, for most of my bank fishing. It's the same rod I use for most of my other fishing on the same size stream, and I'm naturally armed with it when I approach a bank. If I'm on a small stream, I use a rod a foot shorter. On average water, where the obstacles are not tree limbs hanging from overhead, longer rods are an advantage for lifting a backcast over high bankside brush, a disadvantage for wriggling through the same brush.

Bank lies are almost always defined by boulders, indentations in the bank, and an outside bend in the river. The currents push against that bank, erode the greatest depths there, and create somewhat abrupt drop-offs that are essential to form bank lies, like this water on the lower Deschutes River.

The leader should be about the length of the rod, tapered to a tippet that turns the fly over briskly. I seldom go finer than 4X, because I fish where currents are strong, trout often large, and banks so difficult that I can seldom follow a trout when it runs. I want to be able to hold it, turn it, and force it upstream to land it. I can't do that when the trout weighs a couple pounds and my tippet is in the same weight class.

Fishing the Bottom

Fishing the bottom along the banks can be effective at the edges of runs or wherever you want to probe a few pockets near shore without wading into them. This kind of fishing should be done with medium to large weighted nymphs trailed by one small one, rigged under an indicator and lobbed with the kind of gear appropriate for them. You should move slowly, either staying dry-shod on the bank or wading so near to the bank that you can reach out and grab it if you start to tip over into deeper water.

It's best to fish nymphs along the bottom with upstream casts. You'll find occasional places where you can only enter through the brush to an access upstream from a likely holding lie. In such cases, make a short cast downstream and then feed line into the drift of the nymph so that it tumbles as naturally as possible. You'll rarely get a perfect drift, but you'll often get a good whack anyway, because this kind of bank water is so difficult that few folks ever fish it.

Fishing the Mid-depths

Fishing the mid-depths is not particularly productive along banks. It's usually better to fish either on the bottom with a nymph or on the top with a dry fly. I suppose that fishing a dropper off the dry, on two or three feet of tippet, could constitute fishing the mid-depths. I do it often and catch many trout on the dropper when I do. But I don't want to complicate matters by considering it anything but offering the trout a choice by showing them two types of flies on the same cast.

I've found a few banks in streams where it's possible to wade out into the center currents and then fish back in toward the edges. At rare times, a wet fly or soft-hackle, placed tight in and allowed to sink slowly along the bank, will take trout when a nymph or dry fly fails. But it's not a situation you're going to encounter often enough to be any more than minorly aware of it. I can't remember the last time I employed it.

Fishing the Surface

Trout food along banks is largely surface fare, and trout normally hold there with eyes upward, prepared to jump toward the top for a natural or a dry fly. It's by far the most pleasant way to fish along banks. It's also very often the most productive.

An Elk Hair Caddis or Deer Hair Caddis in size 12 to 16 is my usual choice for poking along banks where the water is at all rough, as it is along most banks that I explore with dry flies. These drys work well whenever caddis are out, which is intermittently all season long. I also find that an Elk Hair–style dry works well enough when small stoneflies are out, immature hoppers are hopping, or a variety of other minor terrestrials go plopping to the water. Other good exploratory bank dressings include the Royal Wulff in sizes 12 and 14 and the Stimulator in size 8 to 12. In late summer and early fall, it's difficult to find bank water where a size 10 or 12 Parachute Hopper will be refused. If you own patterns in which you have more faith than mine, stick with them. Trout will probably like them unless they're onto something particular, such as a beetle or ant, in which case you'll need to match it.

Presentation of dry flies should be upstream and tight against the bank. Given a good boulder or an eddy, bank water might be five feet out. Most of the time, the best water is within a foot from shore. Casts more than three feet from the edge should be to specific lies that you have spotted and suspect might hold a hungry trout. In most situations, cast first to cover the water just inches from the bank, then a foot out, then two feet out, and finally three feet out. Fish any water beyond that as lies, with targeted casts.

Within the bankside zone, you'll see many features that should get special attention. Place your fly close into the opening of any indentation in a bank. Cast to the downstream side, then the outside, and finally upstream from any clump of bunchgrass that hangs over the water. Do the same with any boulder. In all but the fastest water, I take as many trout upstream from bankside boulders as I do downstream from them. Use sidearm casts to tuck dry flies as far as you can up under any willow or alder limbs that sweep to the water. These droop with insects, especially when the largest stoneflies are out. A cast driven up there far enough into darkness can draw an explosive take from a trout that hasn't seen many flies.

After you've covered all the water you can reach from one position, move upstream the length of a cast, and creep as near as you can to the next set of prospective edge lies. Cover each subsequent section of bank water, fishing upstream with short, accurate, and controlled casts.

Fishing Eddies along Banks

Eddies are caused by current deflected outward by a point of land, so they're always formed along the banks and constitute a special kind of bank water. Some eddies are tiny, and you can fish these as you come to

A classic eddy, such as this one on Oregon's Deschutes River, is formed by a point jutting into the main current, causing it swerve outward, then turn back in on itself to fill the vacuum left by the deflected currents. The flow is circular, coming back upstream on the inside edge along the bank and forming a vortex in the center where foam gathers and trout foods are gathered as well. Trout hold and feed in the bank currents, which become prime lies. They move out to the revolving central current to feed but won't hold there unless something attracts them there, making that part of the eddy a feeding lie. Always fish the bank water of an eddy, whether you see trout rising there or not. But scope the vortex carefully for rising or suspended trout before spending a lot of time casting over it.

them in the course of your upstream ambulations. Most eddies are ten to twenty feet or more across, especially where a strong current swings out away from shore. You need to approach them from a different direction.

In all eddies, the current right along the bank, on the inside of the eye of the eddy, is reversed and flows upriver rather than downriver. Trout holding and feeding along the bank face into the current, as always, which means that in eddy situations they are facing *downriver*, rather than upriver as they do along most bank water. They're pointed toward you, and looking at you, if you approach the eddy in the normal course of a day spent fishing upstream along the banks. You could move directly into position to fish them from the lower end of the eddy with downstream presentations.

Excellent bank water is always defined by a combination of current, depth, and obstructions to the same current. In this case, the tree upstream is certain to tumble a fairly constant supply of salmon fly, golden stone, and caddis adults to the water. Any smart trout in that section of water would tuck itself up in the shade and wait patiently for its meals to be delivered. Guide Jon Belozur has spotted a big nose poking out and is pointing it out to Ray North, who hooked it on a size 16 Elk Hair Caddis moments later.

But you would get just one shot at them before you'd be forced to lift your line off the water right over their heads. That would end that.

When you come to any eddy larger than a tiny one, you should back away from the bank, go around to the head of the eddy, and fish the reversed bank currents *upstream*, with the trout facing *downriver*, looking away from you.

Large eddies are traps for all sorts of debris, including both aquatic and terrestrial insects. Trout sometimes see large numbers of a single insect, usually small mayflies or caddisflies, ants or beetles, but also on rare occasions giant salmon flies, golden stones, or grasshoppers. If one thing is abundant, trout feed selectively, often hanging near the surface in pods because territoriality gets suspended in the slow water of eddies. You can see the trout as dark ghosts, as noses lifting out and falling back, or sometimes as splashy rises when they dash to outrun their friends and take something off the surface or just under it. What the trout take might be visible or so small you can't see the insects without binoculars. If the trout are big and are feeding on big stoneflies, you'll hear the takes and will have no doubt about what just died. Such rises might cause you to step back from an eddy in fright.

Trout feeding selectively in eddies often present maddening problems. If you're patient enough, you can work into position without alerting the trout, or you can wait for them to return and start feeding again if you do put them down. You can use an aquarium net or scan the water with binoculars to see what they're picking from the current. Then you can extend your tippet to fine and four feet or so long and tie on a floating imitation. Finally, you can present the dry fly with no more than a flick of your rod tip; anything more will put the trout down. But it's still likely that you'll get almost instant drag in all those seething and conflicting currents in the eddy, and the trout will refuse you.

When trout are feeding selectively in eddies along the banks, you have to apply the tactics of flats. Contrive to place the fly *upstream* but close to a rising trout, with lots of slack in the leader. As the slack feeds out, the fly will drift toward the trout; that nose will reappear to inspect your fly, and it might take the fly under with it when it subsides. That can happen, but it won't happen every time. Most of the time, the fly will be rejected. Success in eddies takes a lot of practice added to a lot of patience.

It also takes a lot of luck. Every eddy has hundreds of almost invisible currents. You can put everything together in ways that you perceive to be perfect, and the whim of a tendril of water can pull it all apart in a second. Trout will perceive drag on your dry fly and refuse it. The solution is usually persistence. Keep casting until the world becomes perfect, and the trout will climb onto the fly.

A second solution is to tie on a tiny soft-hackled wet fly in place of the dry, the same size and color as the floating natural, whether it's a mayfly dun, midge adult, or caddis. It's surprising how often eddy trout seem to be feeding only on the surface but actually are taking just as many of those same insects either before they've emerged or after they've drowned. Drag with a wet fly does not kill your chances. If you're unable to notice takes to the tiny fly, add a tuft of yarn three or four feet up the leader, or a white foam or small balloon indicator. You just might begin taking trout. Soft-hackled wets have often solved eddies for me.

Trout in eddies can be surprisingly difficult. They might make you want to stomp off and fish a riffle. Take your time; calm down; read them right; fish them fine; use a dry or switch to sunk. You'll figure them out, though rarely right away.

Fishing Banks from a Boat

I'll lay down just a few rules here and let you learn the rest from your guide, who will want to yell at you if you screw up but won't, because his reputation and therefore his livelihood depend on your catching trout and enjoying it, and you might catch trout but you won't enjoy it if he's yelling at you while you catch them.

The first rule is this: Don't ever lift your eyes from your floating fly to look at the magnificent scenery. If you look away, you're bound to miss a strike.

Fishing Montana's beautiful Bitterroot or broad Yellowstone can cause confusion. You want to look at all the mountains and cottonwoods, watch the tall clouds float by, see the deer and the moose, the geese and the otters. But you want to catch a four- or five-pound brown trout too. Sometimes you wish you could row the boat, so the guide could stare at his fly while you enjoy the scenery. But you try to work out a balance, following the float of your fly, flicking your eyes to read the water ahead, and glancing at the scenery rarely and only briefly, a bit frightened you might miss something when you do.

This leads to one of the rewards for learning to recognize empty water: When you see some coming, steal a glance at Montana and hope you've read it right. If a trout arises out of that "empty water" and goes down with your fly aboard, and you fail to see it happen and set the hook, your guide is going to shout. You can't blame me; I told you not to look away.

Reading bank water becomes doubly important when you fish it from a boat. You want to high-grade prime lies with your casts, but you don't have a lot of time to analyze those lies, and you have to hit the best of them with quick and accurate casts or you'll capture the worst trout.

Recall the three factors of good bank water: *depth*, *current*, and *shelter*. Wherever you see these three things converging as the boat rushes near, plunk your fly down quickly. If nothing comes to it in a hurry, get it out of there and settle it in the next place where all three things appear to come together. A drift boat moves so fast that you'll probably hit every third piece of bank water that you recognize as a prime lie. But that's all right; if you can read the water right, you can hit your share of those lies and catch your share of the trout they hold. Some will be big ones.

If you can't read the water, you'll have to rely on your guide to tell you where to cast. It's a little like having somebody else aim your rifle for you. Firing at random is not as dangerous when your weapon is a fly rod. There's a lot more bad bank water than there is good bank water, but you'll never go home empty if you keep casting at it.

The second rule: When you fish dry flies from a moving boat, cast at an angle ahead of the boat, whether you're in the bow or the stern. The boat, farther out in the stream, is in water that's faster than where the fly is adrift, in close to the bank, where friction slows the current. As the boat overtakes the fly, slack is inserted into your line, and you get a longer drag-free drift. If you cast at an angle behind the boat, the line draws tight, and you get drag after a very short drift.

When fishing banks with streamers or nymphs from a moving boat, you can follow one of two theories, but be sure that you agree with the one to which your guide subscribes. I've fished with guides who were certain about opposite views, and it turns out both were right. The third and final rule: Always agree with your guide, even if you don't.

The first theory is to cast your nymph or streamer at an angle behind the boat. Let it sink until the line begins to draw tight, because the boat is in faster water than the fly. Then start to strip, retrieve for a few feet, lift the fly, and hit the next bit of bank water at an angle behind the boat. The tight line lets you feel a strike and get the hook into the trout. The worst sin when boat fishing with sunk flies is letting a lot of slack get into your line so you can't tell when you have a take.

The second theory is to cast at the same angle ahead of the boat that you would with a dry fly. Give the nymph or streamer a few seconds to sink, during which time you should slowly strip line to prevent slack from piling up. When you feel the fly is at the depth you want, which need not be deep, begin retrieving. You'll have to strip much more vigorously than you would when casting behind the boat because you need to compensate for the slack forming as the boat catches up with the fly. The line will belly out from the bank, and your retrieve will draw the nymph or streamer toward the center of the stream. Retrieve a few feet, lift the fly, and cast again.

I've used both methods, and both work. On a narrow river, where the passing boat might be noticed by trout holding on the banks, the downstream method is better. On wider rivers, I tend to prefer the cast behind the boat. It lets me be a bit lazier, and I seem to catch as many trout. You'll have to choose, or let your guide choose for you.

Whether you're fishing from a boat or the bank, always remember that bank water is *tight against the bank*, not five feet away from it. Keep your flies in those prime lies inches to a foot or two out, and you'll take trout from banks, so long as you're fishing bank water that holds trout.

Special Techniques for Mountain Creeks

Mountain creeks are special places for me. I grew up fishing forested hills along the coast of Oregon. Tiny streams bound whitely down to the Pacific Ocean. Wherever waterfalls block upstream migrations of salmon and steelhead, pools in the secret headwaters above hide bands of bold and hungry cutthroat trout.

Creeks offer isolation. More important to an angler seeking lessons in reading trout water, they offer an intimacy that you don't get on larger streams and rivers. Small creeks have an abundance of trout, and though the trout are small like the streams in which they live, those trout are always willing to tell you if you've learned your lessons well.

The type of terrain dictates the nature of a stream. Most mountain creeks have steep gradients, tumbling swiftly until they gather others of their kind, slow a bit, and become the larger foothill trout streams and small rivers that we fish most of the time. But most mountain creeks also have gentler stretches, mellowing a bit for a hundred yards, or even half a mile or more. A few have meadow stream reaches where they meander, form undercut banks, and look like spring creeks.

The structure of a streambed is dictated by the nature of the stream. The higher into the headwaters of any river system you go, the less the effects of erosion, and the more likely the rocks will be large. On my favorite coastal mountain creek, I'm constantly forced to climb boulders bigger than the pickup that delivered me over the old logging roads to the lip of the canyon that's as near as any road comes to the distant creek bottom.

Important holding lies in mountain creeks are almost always plunge pools, short runs, brief tailouts of runs and pools, or pocket water lies amid cascades. Water types that are relatively rare include riffles, flats, and bank water, though on the narrowest creeks, using the widest definitions, it's almost all bank water.

A typical mountain stream pool is formed where the current, bounding over boulders, has gouged out enough of a deep spot in the streambed for the water to pool up. In this situation, on an Oregon creek so small I'd be foolish to be able to remember its name, Masako is dapping a dry fly onto all the water that is slow enough for its surface to be smooth enough that her fly will not be subsumed by whitewater. The most likely prime lie is right at the end of the current tongue, where it is deflected by the half-submerged boulder with the rounded top. That dark slick along the mossy boulder on the left side of the pool would clearly hold a trout, perhaps a nice one. The almost eddied water at the head of the pool, to the right of the current and protected by those branches, is best of all for casters able to get a fly in there. I'd crouch and do it side-armed and left-handed, but then I've spent far too much of my life prowling small streams.

I will state here, however, that taking the time to train your off arm to cast at least reasonably well will open a lot of trout lies to you that have been closed in your past. I'll also admit that it takes a lot of time to learn to cast accurately with your off hand.

Few of the needs of trout are met generously on mountain creeks. That's why most trout remain small. The need for shelter is met in plunge pools, benches, and runs. Long stretches of cascading water provide only an occasional boulder or downed log to break the current enough for a trout to

Many mountain creeks, in this case the upper Tongue River in Wyoming, have meadow stream reaches. This one is open and alpine, rather than lowland or forested. Its pools are very easy to read: Trout would hold in relationship to the main flow and deepest depths, where they would find their needs for shelter from currents, protection from predators, and food all met. Plenty of boulders along the bottom, a few of them protruding like those to Jim's left, break the current. He's already taken a trout or two from behind those boulders. The most productive water looks as if it's right up the current tongue, especially on the left under that tree, where it takes a bit of a turn and forms a bit of a corner. That is prime.

get out of the flow. Winter and spring brutality in a mountain creek, when the water is high, reduces the number of sheltering places to a very few and might be the limiting factor to the number of trout that survive through spate flows until summer.

The need for protection from predators is satisfied to a certain extent by the many large boulders and ledges beneath which trout can hide. But when they come out to feed, usually in shallow water, they're exposed. When water levels reach summer lows, hiding places are reduced to a very few. The nearness of overhanging branches on a narrow waterway increases the danger from patient kingfishers. I never fish my favorite streams without startling, and being startled by, one or two of these raucous hunters.

The need for food is met sparsely on most mountain creeks. That is one of the reasons their trout are so eager for your flies: They're poorly fed and always hungry.

Foods in such streams are extremely varied. Most aquatic insect types have adapted to niches within creeks. There are swimming and crawling and clinging mayfly nymphs, though almost no burrowers. Stoneflies and caddisflies are abundant. Midges are replaced by blackflies. Such variety is a sign of the health of any stream. Diversity of insect species reports a healthy environment; a narrow list of species tells you something is wrong, and that we likely did it, though it's not always deleterious to trout, or to trout fishing, as tailwaters prove.

All this variety makes it rare that any one insect is dominant in a mountain creek. That's why you seldom match a hatch on one. An appropriate attractor dressing will almost always incite trout to strike.

Terrestrial insects are important in relation to the width of a small stream. The narrower it is, the higher the proportion of terrestrials that make their way into the stream's drift. Creeks, with their overshadowing conifers and alders, are rich in landborne insects: beetles, carpenter ants, termites, leafhoppers, and inchworms. Trout see such a wide variety of them that they rarely become selective to any one kind.

Territoriality is an important factor on the smallest streams. Prime holding lies are few and usually small. The size of a territory dictates the amount of food a trout will see and is one of the major factors limiting the size of trout caught on small water. If you catch a trout that stands above the others in size, the size of the lie from which you caught it will also stand out above the rest of those in the stream. That is one reason why reading water is so important on mountain creeks. It also provides a way you can learn more about reading trout water: If you begin to notice what kind of water gives up the largest trout in a small stream, you'll quickly learn to notice where the three needs of trout are met in one place, forming a prime lie, on bigger water.

Trout in creeks usually hold on stations, watching territories for drift. If conditions are such that a fair amount of feed is being delivered to them, they rest on cocked fins, ready to race for any promising bite that lands on the water near them. They feed on the drift at all levels, since all the levels, from the bottom through the mid-depths to the top, are compacted together in shallow mountain waters.

"Fish those current tongues," my father constantly advised me when I first began fishing tiny streams. "That's where trout hold, right under the current tongues. Ignore the rest of the water." For years I did as I was told, fishing only the frothed water where cascades or tiny waterfalls plunged

It took me a long time to realize that you should read an entire pool and fish it all, rather than just rushing up to stand at the foot of it and casting to the current tongue at the head of it. In this case, on Big Indian Creek in the mountains of eastern Oregon, there are potential lies on both sides of that tailout, as well as the obvious lies where the current comes in from the slight plunge upstream.

into pools. I caught lots of trout, more and more of them as I learned how to cast to the water instead of the trees, how to tie dry flies that floated for more than a few moments, how to read current tongues and fish them better. But it began to intrigue me that I saw so many dark ghosts of trout fleeing up from tailouts as soon as I entered pools.

Finally I ignored the advice and began taking a few preliminary casts at the tailout of each pool before I poked my head up to where any trout holding there could see me. This was frustrating fishing at first. It's easy to stand at the foot of a short pool, cast to its head, and get a good float on its current tongue. All you have to do is pop your fly to the water, and the fly will drift naturally back downstream toward you.

A tailout is another matter entirely. The water is shallow, slow, and clear. Trout holding in it are spooky. Your cast must go over their heads. The current gathers to drop into the rush of faster water downstream. Your fly lands where the water creeps, but your line lands on that swiftness and almost immediately you have drag. Even a trout in a mountain creek knows enough to bolt when a fly begins to speedboat across the top of its shallow tailout.

After some time, I learned to crouch low and make my casts upstream at an angle from off to the side of a tailout so that most of my line landed on the bank, with only the leader and fly on the water. Or I learned to hook my casts so the fly came in where I wanted it but the line looped off to the side and ended up on land. If a boulder protruded from any tailout, I learned to place my fly and leader upstream from it, while draping the line over it. This prevented drag, and I would often get a trout. As you can see, I fish truly narrow creeks.

I was astonished to learn how many large trout held in the lower parts of pools, and as the season wore on and the water dropped lower and lower, more and more good trout moved down from current tongues to hold on tailouts and in the still parts of pools. The reason is evident in hindsight: In spring, most of the feed is aquatic insects, which arrive from the fast water upstream, and the prime lie for a trout is at the head of a pool. But as the season heats up, most feed becomes terrestrial life dropped from above, and the prime lie now is the point from which a trout can keep an eye on the largest part of the surface of the pool: the tailout.

My father, who advised me to fish just the current tongues, fished mountain creeks only in June and July, when their forested watersheds held lots of groundwater and gave it up slowly, and streams held good flows. After that, he considered streams to be unfishably low and quit. I crouch lower now, as the season goes on, and take far more than half of my trout from tailouts in August and September.

The body of a small mountain pool is often no more than three or four feet deep. But it's always deeper than the water around it. The forest canopy is dense, unless the stream has been logged, in which case it's finished and it's time to find another. The main current often butts against a cliff or steep bank, putting it in shadow, which makes it a lot darker. Darkness is to me the definition of some of the best water in the smallest streams.

It took years for me to learn to cast a dry fly over black pools that were almost still waters, then let it sit while I thought about something else and trout watched it, perhaps thinking about something else as well. For some reason, a fly can sit unattended for a minute or two, in full sight of whatever trout might be down there. Then suddenly something stimulates one of the trout—anger or curiosity or hunger or impatience; I still don't know—and the pool is rent.

I continue to fish current tongues at the heads of pools, just as I always did. And I take lots of trout there, drifting a dry fly or nymph on the frothed water or just to the sides of it, down the seams where fast water and slow

A Sense for the Seasons

When you're reading small-stream pools, keep in mind that shifts in the foods that trout find available cause shifts in where they can be found in the pools. In spring and early summer, the most consistent sources are aquatic insect hatches, such as the caddis shown here. At that time of year, trout are rewarded by holding near the heads of pools, where mayflies, stoneflies, and caddisflies—nymphs and larvae and pupae and adults—are all delivered down the currents. In late summer and fall, however, when aquatic insect hatches taper off but terrestrials such as this beetle begin roaming around, a greater abundance of trout foods fall to the broader middle and lower parts of pools from trees or streamside shrubs and grasses. Trout are better fed if they back down lower in pools, where they have a better view for things plopping in from above.

I've already fished the thin tailout of this pool and took two or three trout from water that should never hold them. One came on the flat just off the rocks to the right side of the pool; the rest came from the slick along that left bank. Now I'm preparing to move up and fish the incoming current tongue, which in my youth would have been the first and only place I fished in the pool.

are sewn together. But now I fish all the parts of a pool—tailout, body, and head—and I take nearly equal numbers of trout from each one.

Tackle for this kind of fishing should suit the size of the stream. This does not mean going down to the very lightest 1- and 2-weight outfits. Those are soft, for presenting tiny flies with open loops. Dry flies for small streams are usually tied on size 12 to 16 hooks and are often heavily hackled for flotation. It takes a bit of line weight to turn them over and cast them with authority.

I prefer a 7-foot rod for a 4-weight line, though a 7½-footer for a 3- to 5-weight would be at least as effective. In most of my early fishing, and a lot of what I still do today, branches overhead are a more common obstacle than brush behind. A shorter rod keeps my line lower and lets me fish a lot more water than a longer rod might. For me, the primary criteria for a rod on a mountain creek are quickness and accuracy. I don't like a weepy rod for short, tight casts with a dry fly. A soft rod that does not direct the fly exactly where I want it frustrates me when I'm trying to make twenty- to thirty-foot shots at targets smaller than a basketball hoop, usually with a sidearm cast.

I use double-taper floating lines on small streams. If your rod is better balanced with a weight-forward taper, it doesn't make much difference, since casts will be too short to involve the full belly of any line of that type. The leader should be about the length of the rod, with a two-foot tippet added in midsummer when the water gets low. I use 4X and 5X tippets. They're a little light for bushy dry flies, but with short leaders they turn over well enough, and anything stouter would seem a shame matched against ten- to twelve-inch trout.

Dry flies for mountain streams are usually attractors. The Royal Wulff is excellent. For several seasons, I used nothing else. I could see it because of its white wings, and trout took it eagerly. They still do. At the advent of the Elk Hair Caddis, I switched to it and used it almost exclusively for the next few years. Through subsequent seasons, I've added quite a few flies to the single fly box I still carry on my home creeks. Now it contains nymphs, wet flies, and even a couple streamers.

The nymphs are generic. Yours would be as good as mine. I use my favorite Beadhead Fox Squirrel in size 16 because it works on all waters, an olive beadhead in size 12 because it looks like an inchworm, and an A. P. Black in size 14 because I tried it once and trout liked it that day and have continued to like it ever since. The streamers are the Muddler in size 6 because when I used to kill and eat small-stream trout, I found an outsize number of them had dined on sculpins, and a size 10 olive lead-eyed Woolly Bugger because I once discovered that such a horror smacked to the water would often incite an attack rather than an instinct to flee. I don't know why and don't want to know.

One wet fly I've added is the Partridge and Yellow soft-hackle. I played with one once when the sun was bright and trout refused to come to the surface for dry flies. They surrounded that soft-hackle almost in clouds, fighting each other for it. I could see it adrift in shallow, clear water when cast close enough. It was fun. It still is fun.

Successful tactics on small streams have three parts. First, fish the tailout, body, and head of each pool or short run as you come to it, not skipping over one to get to the next, because you'll frighten trout out of one type of water, and they in turn will rush up and frighten trout out of all the other types. If you fish for a couple hours and discover that trout are holding in only one type of water, then it's fine to concentrate your casting there.

Second, stalk the water, not the trout. Learn to read potential lies, then assume each of them holds trout. Likely it will. Cast over it as if it does, and you will have read the water by exploring it with your fly.

Third, present your fly so it lands on the water rather abruptly, as if it were a natural insect surprised to find itself there. Trout will be surprised too, and they'll move to take it before they have time to decide it's the wrong kind of groceries. I don't mean you should smack the fly to the water, but don't try to achieve the delicate presentations on mountain creeks that you do on spring creek flats. Briskly may be a better word. Trout expect their food to arrive under protest about being on water. Your fly should have that same attitude, if you can manage to impart it. A brisk outfit will help you.

A full day on a mountain creek usually has a specific shape for me. It starts out in the morning with a bright dry fly. Light is a bit subdued before the sun rises to where it shines down into the canyon and strikes the water. A visible dry is easy to follow against the darkness of the stream. And trout are almost always willing to take dry flies early in the day, perhaps because fare is slight until the sun warms natural insects and the fish are hungry.

When the sun gets up in its arc, around ten or eleven o'clock, so that it strikes straight down, I often switch to a soft-hackled wet fly, the Partridge and Yellow in size 10 or 12. Trout are often reluctant to rise to dry flies in bright sunlight, but they move dashingly for anything that comes to them in the drift, beneath the surface. At times, in bright light, I take two or three times as many trout on wets than I can draw up to drys.

Upstream wet-fly fishing requires intense concentration. Nothing will put you in closer touch with a tiny stream. You have to wade so near that you're casting little more than the leader, and then watch for any twitch of

Trout in typical dark mountain creek pools are not the most wary in the world, because they're far less pestered than trout living in waters that are easier to fish. But you should always honor their senses: Don't stomp up to them, sending vibrations through the water to warn them; don't thrash around with your rod or send your line over their heads; don't loom into their sight. Robert Gorman demonstrates the proper approach here, on his knees on a nameless little Oregon stream. He is owner of Green Mountain Rods; his 7-foot, 3-weight is one of my favorite friends on my home creeks.

The prime lie in this pool is right where the two current tongues join, therefore offering trout food from both directions, along with what is probably the deepest part of the pool. But both current tongues, and also the still, black corner lies on the two sides of each, would also hold trout. With the cautious approach Robert made here, he was able to extract trout from several parts of the pool.

your line tip or leader butt or the subdued wink underwater that tells you a trout has taken your fly. After you've fished this way for a season or so, you begin to take trout without any knowledge of what made you raise the rod to set the hook. But a trout is always there.

When the sun has tipped over the canyon in the afternoon and no longer strikes the water directly, I find that lots of caddis adults, mayfly spinners, and small stoneflies come out and begin the dances that will last until evening. I switch back to a dry fly, usually the size 14 Elk Hair Caddis that at one time was the only fly I used on tiny streams. As light fades, trout take it more and more willingly.

If trout refuse any of these fly types during any part of the day, on any of the water types, I switch to a nymph beneath a yellow yarn indicator, dangled on a couple feet of 4X or 5X tippet. It almost always works. It would almost always work even when one of the other methods is working as well. I use it only as a fallback, though I suppose I could use it first and try all the other methods if it failed. I suspect, however, that it's best to keep it in reserve, because if the nymph failed, all else probably would too.

I use the streamers in very rare instances where I want to sweep the Muddler over a tailout or plunge the lead-eyed Woolly Bugger into the depths of a pool that has refused to surrender any trout. Sometimes this works. Usually it doesn't. I always feel guilty when I try it.

The three-part approach to the water, fishing the tailout, body, and head of every run and pool, has added a lot of trout to my small-stream fishing, many more than I used to catch when I fished only current tongues. The three-part approach to the day, changing types of flies with the waxing and waning of the sun, has changed the hours when I expected only an occasional trout into hours when I expect a trout from almost every holding lie. On rainy days, when trout aren't always eager for dry flies during any hour, the fallback nymph has kept me happy.

My tackle and tactics are under constant refinement. I discover something new almost every time out. If I fished your mountain streams, I might find that new flies and tactics would work better there. Not long ago, I fished for a week on Japanese mountain streams with my wife and Migaku Saito. His only fly was a size 12 parachute that I came to call the Saito-San Special. It was devastating in Japan. I brought it back and tried it on my old home streams. I was astonished to find that it outfished the flies I'd always used on them. The fly has a rusty Superfine body, dun tails and parachute hackle, and a lamb's wool wing post. That's all there is to it. I now tie it and use it on mountain creeks more than any other dry fly.

Creeks are clinics when it comes to the study of finding trout and discovering what works to take them. They're often demeaned by those who fish larger waters for larger trout. But they can give you lots of lessons that prepare you for those other fishing situations. Be careful; you might find that you come to prefer mountain streams.

12

Classic Freestone Trout Streams

Medium-size trout streams are what most of us fish most of the time. They're classic water, a personable size, usually not difficult to wade. They average about a long fly cast across, comfortable to cover. They're the kind of water that a major percentage of this book has been about, so they'll get just a minor review here.

Classic trout streams are freestone, arising from runoff and groundwater sources. They have all the types of water and holding lies discussed earlier in the book. They display the standard riffle-to-run-to-pool structure. They get knocked out of shape by rains in winter, by snowmelt in spring. Many are low and clear and fishable, even offering hatches and rising trout, in late winter and early spring before runoff begins. Most are unfishable for a few weeks while watersheds clear out their accumulations of snow, if they're in the latitudes that get snowpack. They're perfect for fishing after those freshets subside, any time from April through July . . . that last month is spring on the high Yellowstone Plateau. They're subject to low flows and often have temperature and oxygen problems, some as early as late June, most in August and early September. They're refreshed again for a few weeks in fall, and most people turn their backs on what can be a couple months of very good fishing after the first early frosts of October.

These classic streams are deep enough in most of their stretches, for most of the year, to make it profitable to keep in mind the three levels of the water: bottom, mid-depths, and top. Each type of water, and each level within each type, is important at different times, according to where and when trout hold there and whether they rest or feed actively when they're there.

The structure of any trout stream depends on the geography through which it flows. Its gradient reflects the steepness of the country around it, and its streambed is determined largely by its gradient. The closer a classic stream lies to its mountain creek sources, the more bouldered its course

Freestone streams are defined by their bottoms, made up of the things that give them their awkward angler's name: free stones. Although *freestone* is a term we use often, it's one you're not likely to find in a scientific book on geology, biology, or hydrology. Still, it describes what we mean, which is a creek, stream, or river that has eroded its bed to a composition of rocks of all sizes, not silt or sand and not solid bedrock.

The typical structure of a freestone stream of any size is riffle-run-pool, though you'll not often find them lined up and readable in such pleasant and predictable order as they are in this photo, on a small tributary to the Elk River in British Columbia. Here the empty riffle upstream bounds over a shelf, forming several short, white current tongues that calm down and become a run along the boulders, then slow and deepen slightly into a very short pool before rising as a tailout and tipping over abruptly into the next riffle below. All of the most likely looking holding lies are down the length of the run, where it pushes against and around the boulders in the center and on the far side. The prime part of this water is from the half-submerged boulder to Masako's front upstream to the pyramid-shaped boulder nearer the head.

All streams, whether freestone or meadow, small or large, mountain or flat-land, are shaped by the geography through which they flow. The upper reaches of Utah's Logan River, shown here, step down off the Rocky Mountains in a series of benches. They're easily read. Trout hold either under or alongside the current tongues that probe between the boulders forming the drops from bench to bench. The current is slower on the benches. The water is deeper on them. The currents break up the surface, hindering a bird's ability to spot trout. Food is delivered down those currents. All three of the primary needs of trout are met in association with those current lines, and that's where you'll find the prime lies.

will be and the closer its fishing will be to the plunge pool type described in the previous chapter.

As a stream grows and approaches downstream maturity, it flattens. Its course tends to meander more, and its bottom is composed of finer and finer material, first predominantly rock, then cobble, then pebbles, and finally sand. The plunge pool nature higher in the headwaters gives way to the riffle-run-pool nature of the foothills and, on many streams, to the meanders of the flat valleys.

The water types that are most important in a given stream also depend on its gradient. In average trout streams, all the water types are present, and most are important. Choppy riffles provide an abundance of feed and some holding lies. Three- to four-foot-deep runs have lots of holding water but a bit less feed. The three parts of pools—head, body, and tailout—can all be important. Rapids and cascades have pockets of fine holding water. The banks along most of a classic freestone trout stream are eroded too flatly and therefore too shallow, and the water is too slow along them or lacks sufficient features, to hold trout. But there will always be some bank water that meets the three needs of trout. Those good banks will be noticed only by observant anglers, who will find them not only very productive, but also rarely fished.

Productive flats with rooted vegetation are almost always absent from classic freestone trout streams. Aquatic plants rarely have a chance to take root. You'll find flats, but they'll be freestone with cobbled bottoms, having few places where the current is broken enough to shelter trout of much size. Flats in classic freestone trout streams are often empty water. If they're not empty, have another gander at them; it's likely you're looking at a riffle.

The needs of trout in medium-size streams are met in almost all the ways that have been discussed throughout this book. These streams are large enough to offer shelter from currents in most of the water types. Unfeatured riffles might be too thin to have more than half a dozen holding lies for trout half a foot long, but others will be studded with boulders, shelves, ledges, and trenches that hold trout scattered throughout. More and often larger trout will hold in runs or pools nearby and move up to the riffles to feed when insects are active enough to make it worth the risk of predation and the loss of energy required to feed in shallow, brisk water.

Protection from predators is a factor that trout respond to in many reaches of classic streams. You'll find empty water that would hold trout if only it met this need or had a bomb shelter nearby to which trout could bolt when threatened by bird, otter, or man. In larger freestone streams, most water types offer sufficient protection until late in summer when the water is low. Trout hold throughout the stream until the water thins, then move into the most sheltered areas as the water drops. Being able to read water as it changes, and realizing that it no longer meets the needs of trout in one place but does so in another, will give you a clue to trout movements as the season progresses.

Movement of trout from one water type to the next is perhaps greatest in medium-size streams, where all the water types might be within a hundred feet of each other. At times you'll find all the trout, or at least all the willing trout, in one type of water. When this happens, you can hop from riffle to riffle or run to run, leaving other anglers to fish all the water, in which case they'll be spending most of their time on water that is not productive at that moment. But water that holds active trout one day, or one hour, might not hold them the next. A hatch might end. The trout might move—they *will* move—leaving you to fish unproductive water while those other guys who were failures earlier begin catching trout. Keep your eyes and mind open, aware of change.

The need for food is met in classic freestone streams by all the variety that both the aquatic and terrestrial worlds have to offer. But there is a greater chance on medium-size trout streams than on mountain creeks that one species of insects will be dominant and that you'll have to cast a fly that resembles it at least reasonably in order to take trout.

As in mountain creeks, the importance of terrestrial insects is related to the size of the stream and the type of vegetation growth along its banks. The narrower the stream, the greater the proportion of bank water and thus the importance of terrestrials. The wider the stream and the more water that lies away from its banks, the less often terrestrials will be important, except right along those banks.

The kinds of terrestrials that are important vary with the type of vegetation along the banks. In grasslands, grasshoppers are most important, but you'll also find beetles and ants dropping in and trout feeding on them selectively at times. In forestlands, beetles, ants, termites, inchworms, and leafhoppers get the most nods from trout. When a specific terrestrial is dominant, a match for it will help you fool trout. When none is dominant, a match for one that is at least present will still help you fool trout.

Trout are occupied most of the time in freestone streams with the need to find food. They hold their stations, winnowing insects from the drift at all levels of their territories. Sometimes they drift a little higher in the water column, remaining on their stations, to feed on something adrift in the mid-depths. At other times they move up to hold and feed just under or on the surface. If a heavy hatch starts, trout might move to a location that enables them to feed more successfully: a feeding lie.

Freestone streams have holding, feeding, and prime lies. When trout rest on a holding lie, they're usually, but not always, willing to feed. When on feeding lies, they're usually feeding on something specific, and you need to fish a fly that looks at least a little like it and present that fly at the right level in the water, with the right motion or lack of it to represent the natural insect's behavior. When trout are on prime lies, they'll be holding a station and feeding on whatever the current brings. They'll usually move for a properly presented fly, but they won't go very far to get it. If they're feeding opportunistically, they'll take generic patterns. If they're focused on one food form being trotted constantly through their prime lies, you need to collect it, get a close look at it, and match it.

Tackle on freestone streams should be about what we consider average or standard trout stream gear. Rods should be $8\frac{1}{2}$ to 9 feet long, balanced to cast 4- to at most 6-weight floating lines. Most people today use weight-

Freestone streams offer all the types of lies that trout prefer, and your ability to read them will help you increase your take whenever you fish them. This long stretch of water on the Logan River is not a pool, but a shallow run with some boulders down its length and a riffle at its head. The shallow water on the near side of the stream, though it's more than half of the surface area, is too thin to hold trout. It's empty water. The somewhat deeper but still exposed slice of water, ten to fifteen feet wide, to Masako's front and extending for a long way upstream, would become a feeding lie if insects began to hatch there. We had that happen: Lesser green drakes (*Drunella coloradensis*) came off in that water in large numbers later, and trout moved out to feed on them there. We became too busy catching them to fiddle with photography.

Holding lies form along that entire far bank, where trout find sufficient depth and cover to be comfortable at all times. Masako has read the one prime lie—in the shade of that tree, where the water is broken by a couple boulders just upstream—and by drifting her dry fly over it, she has managed to draw a trout thrashing out of it.

forward lines for the slight advantage they give in casting distance. I feel that the streams I fish have not grown and still use double-tapers if they suit the rod I'm using. They cast far enough for me and give me a bit more control over the drift of my fly.

Most of the time, my leader for classic stream fishing is about the length of the rod. I buy 7½-foot, 3X base leaders and extend tippets from there, depending on the situation. Since I always have a permanent leader butt on

An Easy One to Read

This is a fairly typical freestone stream situation, on Alexander Creek, a tributary to British Columbia's Elk River, if there is any such thing as a typical scenario on any freestone stream . . . they're all so varied. The chattery riffle at its head would distribute a fairly steady supply of food down the current line that unfortunately drives under those trees. You know a trout lurks under there, but you don't know how far back or what sort of view it keeps on the world outside the trees. But it's impossible to pass up.

I cast a size 14 Royal Wulff to the lower end of the riffle and let it skirt the trees, in danger of tangling in them. Before it got far, however, a trout came out and rescued it. It was a wild cutthroat. I was surprised at its plumpness and size, coming from such a small freestone lie.

my line, about a foot long, the leader is a bit longer to start with. I use it at its original length, or lengthen or shorten it a bit, if the situation calls for deep nymphs. I add a 4X tippet for mid-depth nymph and wet-fly fishing. When I switch to emergers or drys, I add a tippet or tapering sections and then a tippet, depending on the size flies I'll ask the leader to turn over.

Effective presentations on freestone streams are usually upstream; this suits the normal direction of movement, which is also upstream. You can fish most holding lies and prime lies as you encounter them, probing each specific lie as you come to it. Fish each of the water types, and each of its levels, according to the best way to take trout from it. This generally requires a bit of tackle changing, from dry to wet or nymph and back to dry again, as you hike along. But the slight extra effort is worth it, and it will probably cease as soon as you discover the one water type and one level at which trout prefer to feed that day.

On a typical day astream, I start fishing upstream, planning to move fast to cover quite a bit of water early, to make sure I see and sample all the different water types the stream offers. If the stream is big enough that I'm boating it, the float gives me a view of all that the water offers. But boating is usually confined to small or large rivers and is rarely useful on freestone streams of the size that we refer to as trout streams.

Avoid the mistake of going to a stream, starting in one likely looking place, and remaining rooted to that one spot all day, or even all of an hour. I don't know how many times I've seen folks fishing fruitlessly in one water type, far more gracefully than I ever could, not catching a thing but never moving, while I flailed away on my hectic rambles along the stream and took far more trout because they happened to be more active just a little distance up or down the stream. Even if they're not more active else-where, once you've pestered a bit of water for more than fifteen minutes or so, you've worn it out. Take another position in that water or move out of it to another type of water. Movement is an essential part of any trout-fishing strategy. If you're still catching trout, though, never walk away from them. That's a more obvious essential element.

On a typical day on a typical freestone trout stream, I'll start with a searching dry fly if winds and weathers promise that trout might be willing to come to the surface. In the absence of any specific insect, I'll try the normal run of Elk Hair Caddis or Deer Hair Caddis in size 12 to 16. I'll fish riffles and runs, especially riffle corners and any visible boulder lies in runs. These are among the most likely places to find trout. If the dry fails to produce in these types of water, I'll drop a small nymph off the hook bend and try that for a while in the same water types, or in whatever water type I arrive at in my travels.

If no trout come to either the dry fly or dropped nymph in a reasonable amount of time, from half an hour to an hour, which is probably too long, I'll switch to fishing a nymph down on the bottom. If winds and weathers appear unsuitable for dry flies in the first place, I'll go directly to the nymph, or more likely a pair of them, without any stops to try anything else. I prefer to fish with uncomplicated gear if it offers a reasonable chance to catch trout, so I try drys when I shouldn't and stick with them longer than I should.

Always keep an eye out for rising trout. If you find them, rig appropriately to fish for them. I've fished a few times with folks who pass up rising trout, usually saying, "Let's go upstream. I know a place. There's better water up there." Off they trot. There is always better water up there, but most times there are no rising trout in it.

Never pass up rising trout is my motto.

13

Tackle and Tactics for Large Freestone Rivers

Large rivers can seem daunting, especially if you're used to fishing riffles, runs, and pools in smaller streams. But they're not nearly as difficult as they might appear at first. The secret is simply to break big rivers into their parts—those same riffles, runs, pools, pockets, flats, and productive bits of bank water that you fish on all types of trout streams. Read each for the way it meets the needs of trout. Fish each according to its kind.

As with all trout water, the type of terrain through which a large river flows determines its gradient, and its gradient dictates its structure. Large rivers are at the lower ends of watersheds, where they've gathered first-order streams, mountain creeks, even a few typical trout streams and have usually reached some sort of maturity. Most large rivers flow through broad valleys. Erosion has had many millions of years to flatten out the riverbed and grind boulders down to stones, pebbles, and even sand or silt. Riffles and runs are common, rapids and cascades rare. If you find pools, they're often so large they amount almost to ponds.

A few big rivers are located in mountainous regions and have more of the features of midsize streams, fewer of the structures typical for large rivers. The upper Madison River in Montana, as an example, is called a fifty-mile riffle. It's big, with a broad valley. But its gradient is fairly steep, and it has lots of boulders and pocket water lies. It flows in and out of a lake formed by a landslide; those somewhat stabilized flows downstream create many miles of excellent bank water. There are few of the large and deep interruptions that you might define as pools.

The Yellowstone River has a more classic structure in most of its miles, but it tips down through lava flows in Yankee Jim Canyon, at the upper end of Paradise Valley. The resulting rush has more difficult water and less riffle, run, and pool structure. The Big Hole River has eroded its way more recently through some serious mountains. It has pools so wide and deep you can't cast far enough to cover them with a fly, unless you're afloat in a

The best way to read a big river is to break it into its parts, then read each of those parts for its most promising lies. On this broad slice of the Bighorn River in Montana, these anglers have wisely parked their boat in the most likely piece of water in the whole photo. A shallow riffle chatters over the water around that tiny island on which they've gotten out to fish. The riffle feeds into a well-defined run just downstream, no doubt delivering lots of insects with it. The long current seam cutting across in front of them would be a prime lie, always holding trout and promising rising fish any time a hatch might be going on, which on the Bighorn River would be often.

Across the river, the channel cutting between the big island in midstream and the small island to its right would also be prime habitat. It's difficult to read it from this distance, but you should anchor your boat there, get out to take a closer look, and read that part of the river just as you would the midsize stream it resembles.

boat, which you are most likely going to be on that big river. The Big Hole River is often described as the world's best example of a classic trout stream. But the sizes of its various water types are greatly enlarged.

My home Deschutes is a big river, but it's not mature, and many sections of it are still brutal. It used to flow east out of desert flatlands, then was redirected north by the uplift of the Cascade Mountains less than four million years ago. That is recent in geologic time. Lava flows crossed its course in some places even later than that. Despite its size, it's a young river and has some features of a mountain creek, though they're written large.

The Madison River is said to be one long riffle, and if you get out among it, it's easy to read like any other riffle. Trout will hang in water where the fast current is broken by boulders. Because its flows are stabilized by Quake Lake, it has less erosion along its edges than a spate stream might and therefore offers long stretches of great bank water. You can read it all for likely lies much as you would a small stream.

Big rivers have all the various water types that other streams have, and they're also forced to obey the laws of hydraulics. Thus, for example, a big river has riffles every five to seven river widths along its course. These are similar to riffles in smaller water, but they are much more extensive and often so productive that hundreds of trout might live on a riffle's bounty. As with all streams, if the riffle is deep and slow enough and has sufficient interruptions to its forceful current, it will hold trout. If trout cannot hold in the riffle, they'll be found holding in the nearest comfortable water downstream, living on the richness delivered from it.

Runs are what we think about when we think about holding water in big rivers. They usually are wide, with shallow gravel bars on one side, and so deep in the center and toward the other side that the structure of the bottom seldom reflects up to the surface. The best holding water is typically in the deepest part of the run, which is toward the center if the flow is straight or on the outside if the run is located on a bend in the river.

Productive runs in big rivers have a current that is at least definable and usually quite strong. They're generally at least three to four feet deep, but often much deeper, down to eight or ten feet. Deeper than that and it's better to consider them pools and fish them in ways that work in that kind of water. In shallower runs, you can expect to find the same kind of trout you would find at the same depth in smaller streams. In runs typical of larger rivers, the depths you'll need to reach will be farther down, and the trout you expect to catch will be bigger than those you would expect to find in a typical trout stream.

Pools in big rivers are broad, deep, and difficult to fish, and they have the potential for producing the biggest trout in any system. Big pools are almost a study in themselves, like small lakes. And like small lakes, they hold their trout in the deepest water most of the time but send them out to feed at the head, on the tailout, and in the shallows on daily occasion, usually during a hatch or in low light.

Flats are not common features of big freestone rivers. Waters with spring creek origins, such as the famous Henrys Fork of the Snake, are known for their broad, shallow, and almost still flats. Tailwaters are almost defined by productive flats. But spring creeks and tailwaters are the subjects of the next couple chapters.

Pocket water is not a feature we think of when we think of large rivers. Most are too far down their systems and too placid. But some rivers carve through canyons and have awesome rapids and cascades. I backpacked through the Black Canyon of the Yellowstone River some years ago, with my nephew Trevor Hughes. We fished riffles, runs, and pools, as well as some rare touchable pocket water. But we also stood on the trail and looked down at lots of violent water that had never been fished, and never will be. It had many pockets but none that could be reached.

In big rivers, pocket water is normally unwadable. You can fish only the few pockets you can cast to from the banks. If you have any doubt about this kind of water, stay out of it. The combination of powerful currents, uneven bouldered bottoms, and the slipperiness that is typical of a big river's rocks make them excellent places to drown, which is one of the worst things you can accomplish during a day spent trout fishing.

Good bank water is common on large rivers. Many reaches are deep, with mild to strong currents rubbing right against the shore. This becomes especially important water during the migration of giant salmon fly nymphs, when trout wait along the banks for the emerging nymphs, and during the hatch, when the clumsy adults creep around in riverside grasses and willows, falling into the water at irregular intervals that keep big trout waiting, ready to rush upward. Not every river has salmon flies, but all have caddis, which love the edges as well, and all have terrestrials that migrate to the banks, only to find they can go no farther without getting wet feet.

Many big rivers have sections with all the aspects of small to medium trout streams. These are broken up by braided channels or side channels that split the river into more manageable—and more readable—portions. These backwaters, streams within large rivers, should be treated exactly as they appear. They're some of the easiest water to read on a large river, and they provide some of its best fishing.

Whenever you read big water to locate big trout, look for the same things you look for on smaller water: places where the water meets the three needs of the trout and are therefore prime lies.

Rivers have lots of lies where trout can find shelter from strong currents, but you cannot always see signs of them. They're deep in the darkness of a pool or so far down in a run that they're not visible from the surface. When you *can* see them, you should fish them. But most of the time on big water, you'll have to identify likely holding water, then fish all of it to get your fly into the specific prime holding lies.

Protection from predators is almost universal in large rivers, except when trout come out to feed on flats or move up high in the water to feed near the surface of runs or back down onto tailouts of pools. Then they're just as exposed to predation as if they were in the thinnest of mountain streams, and they'll be just as wary.

The primary sources of food in large rivers are the same aquatic and terrestrial insects that make up the trout diet in other streams. The variety of species is likely to be a bit narrower. In large areas of similar water type, such as a long, deep run, a single species is more likely to be dominant. If

When you get an overview of a big river, such as the South Fork of the Snake in Idaho, it's easy to see that it has all the same parts as a small to medium-size trout stream. Some of them are writ large, but others, such as the many riffle corners you can see in this photo, would be no more difficult to fish than they would be on a much smaller river. If I were floating this stretch, which I've done only once, I'd be pulling my drift boat onto those islands and stepping out to fish the corners almost all of them form.

This big riffle corner, feeding out into a long run on Pine Creek in Pennsylvania, is an example of breaking a big river into its parts. The water, when we fished it, was productive from the corner down for several hundred feet. When a sporadic emergence of sulphur mayflies in late afternoon grew into a steady hatch right at evening, it became a vast feeding lie. Trout were not cavorting all over it, but they were rising consistently enough that we could locate them and get a hook into one occasionally.

We read the water first by choosing the most likely location: the riffle corner staked out by the two fellows upstream and the run descending from it. Then we fished the water, with sparse luck, until trout began feeding and we were able to pinpoint them in that elemental first step in reading water: looking for rises.

trout are found feeding on a hatch, they'll be just as selective as they are in any other stream type. You'll have to determine what stage of the insect they take and where they take it: bottom, mid-depths, or top. Then you'll have to select a specific dressing that matches the insect and present it in the manner of the natural's movement.

Terrestrial insects are less likely to be an important food source on a consistent basis to all the trout in a big river, because the water along the banks is a much smaller proportion of the total width of the river. But terrestrial insects are extremely important to trout along those banks, and banks are about the most important bit of water to the boating angler.

The Deschutes is a big, brawling river, but it's not difficult to locate its riffles, wade in at their corners, cast dry flies or nymphs, and prod a few trout into activity. It's my home river, and a warning is in order. Most people I watch wade right through the part that holds the most trout to go out and fish where it's deeper and they're pretty sure the big ones wait. But that's not the case; the big ones are in that soft water where the riffle delivers an assortment of insects on its brisk currents.

Larger organisms become more important in the deep water of larger rivers. Baitfish and crayfish, baby birds, trout fry and sculpins, leeches, and landborne beasts such as mice can all get chased down by large, predatory trout. Most folks fishing large rivers hope to catch large trout. One way to increase that hope is to fish with flies that represent the largest bites a trout is likely to see.

The occupation of trout in a big river is the same as it is for their friends in smaller streams. They spend most of their time on stations, taking feed from territories. The size of the territory is likely to be much larger, and the features that create suitable territories are likely to be less numerous, more scattered. Trout therefore are spread out in big rivers.

Most of a large trout's time is spent holding, resting, waiting for the right time to chase something worth chasing. But a small or average trout feeds on the drift that comes to it. At times, when feed is abundant and concentrated in a certain water type, trout move off their stations and concentrate at the food source. But the larger the trout, the larger the amount

The Missouri River is a big tailwater and is fished most often by floating, rigging with a brace of nymphs and an indicator, casting them out, letting them dangle along the bottom as the boat plods along. But you can read its parts and get out and fish each according to its kind. This big eddy, formed by the boulder in front of Jim, defined by the foam you can see just in front of him, and protected by that tree, almost always provides action for those who take time to notice it.

of food it takes to entice it to move. The largest trout, those weighing more than five pounds, are caught where they live or hunt, and seldom anywhere else, in freestone rivers. You usually have to go to them, or at least get your fly to them, right down in their territories.

Going to them takes special gear and calls for special tactics. Rods for fishing big rivers should be 9 feet or longer and stout enough to propel 6- to 8-weight lines. Most of your fishing should be done along the bottom. Most of the water you fish can be probed with combination floating-sinking lines. A ten-foot sinking tip might get a fly to the bottom in runs three to four feet deep. But for typical big-river runs, a wet-head line, with thirty feet of fast-sinking line and the rest of the running line floating, will get it down faster and keep it down better. My favorite is a depth-charge line such as the Teeny 300.

Many people who fish the largest rivers, such as the Yellowstone downstream from the park, use shooting-taper systems. These have a thirty-foot section of fast-sinking line backed by a hundred feet of Amnesia. A hundred

This broad and featureless stretch of the Bighorn River gives no hints about where its trout might hold, except for the seam Rick Hafele is nymphing. The eddy behind him might be productive if insects were hatching, at which time it would become a feeding lie. But they're not, and no trout are rising in there right now. Rick is fishing the well-marked seam, formed by current pushing its way around that grassy point. It's the prime lie in all the water you can see.

yards of backing line fills the rest of the reel. In order to make the system work over a wide range of water conditions, without having to lug a battery of reels, just carry a series of shooting tapers in various sink rates, all balanced to the same rod. Loops at the end of the running line and the shooting tapers let you switch in a hurry without changing spools on the reel.

Lines to carry on big rivers might include a floater, a slow-sinker, a fast-sinker, and a superfast-sinking head. The line to choose in a given situation is the one that gets your fly to the bottom in that type of water, which calls for calculating the combined factors of the water's speed and depth.

The leader should be kept short when you use a fast-sinking line, four to six feet. It should be strong enough to withstand a brutal strike and play a large trout in a heavy current. That means a tippet in the 1X to 3X range.

Flies for the banks and shallows should be typical of what you would use in those same types of water on smaller streams. If you need to match the hatch, your gear should balance the flies you will use; don't try to fish size 20 drys with the shooting systems described above. They're for fish-

On big rivers, especially in winter, it's always wise to fish the shallower and slower water to the insides of heavy and brutal currents. Guide and fly shop owner John Smeraglio knows the address of almost every trout along this bit of big Deschutes River edge water, from years of experience fishing it. Your ability to read water for the way it meets the needs of trout, and to fish it according to water and weather conditions, will allow you to gain that sort of experience, and learn to locate the trout yourself.

ing big water with big flies for big fish. For presentation fishing, use the same light outfit on a big river that you would use on an average stream.

Most big-water fishing calls for either large, heavily weighted nymphs or streamers. Nymphs should be tied on size 2 to 6 hooks with 3XL shanks. The weight on them, as they're tied by commercial tiers, is the same diameter as the hook shank and extends in tight wraps the length of it. Don't try to toss such a fly with your 4-weight. Streamers should be as big and as stoutly weighted.

Presentation of big flies on big water, when you're wading rather than tossing them from a boat, calls for casts that are slightly upstream to give the fly a substantial portion of its drift just to get down to the bottom. Let the line tighten against it, then swing the fly around until it's straight downstream. You might want to pulse your rod tip, making the fly appear to swim around in the current along the bottom.

It's common to attempt to cast as far as you can when fishing big water, but it's more productive to keep casts within the distance you can handle with grace, say forty to sixty feet. You'll have more control over the drift of the deep fly and more success setting the hook. Where the water warrants it, and your abilities allow it, wade deep, cast long, and try to cover the deepest water, way out there. If you can cast farther than most anglers are able, you might show your fly to a trout that has never seen one before. A lot of people who fish big rivers continually, with gear chosen for just that job, can cast more than a hundred feet with ease, sometimes into the wall of a wind.

It can be worth the cost of the assemblage of that gear to make those long casts, as well as the time it takes to learn to use it correctly. A trout that has never seen a fly might be a big one, especially on a big river.

Be patient on large rivers. You might not catch as many trout. But your chance of catching the biggest trout of your life is greater here than on any other moving water.

Meadow Streams
and Spring Creeks

Meadow streams and spring creeks are about the most beautiful places where trout abound. Their peaceful flows meander through pastoral landscapes, wilderness meadows, and jack pine forests. Their water is clear and bright, with tails of lush green vegetation waving gracefully in soft currents. Aquatic insect hatches are prolific, and even they seem to include the most beautiful of their kind, as if it wouldn't be right to send something ugly to set sail on the surface of pretty water: early-spring blue-winged olives, midseason sulphurs, dancing and dainty caddis in spring, summer, and fall.

Trout in meadow streams are doubly difficult, because they are wary in the first place and get lots of opportunity to examine your fly and its presentation in the second place. They're wary because clear, slow water exposes them to natural predation, and their environment attracts the most deceitful predator of all: man with his fly rod. The same glassy water that exposes them to danger also gives them ample time to inspect whatever you offer them. Rejection is more common than acceptance.

Meadow stream and spring creek trout are the most selective of all. Meadow streams have gentle gradients. If the land were steep, the streams would be freestone. Though we tend to think of them as one type of water and call them both spring creeks, in fact most meadow streams have freestone stretches. The Metolius River in Oregon is a prime example. It originates in springheads. It flows for many miles from those heads through ponderosa flats—meandering and gentle, with a stable bed composed as much of silt as of stone, trailing rooted plants, and steep banks, many of them undercut. Then the peaceful river suddenly tips down and carves a canyon through volcanic rock. It's brutal in this two-mile reach, with a few pushy benches that can be waded, but with great difficulty, and pieces of pocket water, most of which must be fished from shore. Few fishermen dare it.

Spring creeks are the most beautiful places in the world of trout, but they're also among the most difficult to fish. The clear water makes their trout wary and also lets them see your approach, the flash of your line flying through the air over their heads, and the falseness of the phony flies you try to con them into taking for some real midge or mayfly. I confess to reading such streams as Fall River in Oregon by watching for rising trout. That's what I go to spring creeks to enjoy. If trout are not actively feeding, I might read the water for obvious lies, but I also might get out my collecting net, try to see what aquatic insects and other animals make the stream tick, and attempt to ascertain what might be about to hatch and when.

Below this canyon, the Metolius gentles out again into several more miles of typical meadow stream water. It gathers additional springs and grows to where it's deep enough that you can wade only a few of its shallow flats. Then it tips down again and makes its final miles to the Deschutes River in a rush of whitewater, so fast and frothed that little of it can be fished.

Many meadow streams originate as spring creeks, for the simple reason that it requires a stable flow to create the conditions that define meadow streams. But even the most rambunctious freestone streams mellow out in places where the gradient is not steep. In the many days I spent following mountain creeks almost to their sources, I've found these meadow stretches most often where an ancient lakebed was formed by a slide, filled with water, and silted in, then the blockage blew out and the stream cut a meadow reach through the level, soft-soiled lakebed. The result is a meadow stream reach in a stream that is almost brutal everywhere upstream and down from the old lake. But mountains also have benches, and a stream tumbling off a hillside will level out, becoming more meadow than freestone, when it crosses such a bench.

In these peaceful reaches, however short or long, freestone streams take on many of the characteristics of meadow streams. But they're still subject to spate flows, and they won't take on them all. Without stable flows, the stream will alternately erode its banks, then retreat within them. Even in meandering meadow reaches, a freestone stream will tend to have gravel bars on the insides of bends and deeper bank water only on the outsides. They will always be subject to scour, so they will never harbor attached algae or rooted vegetation.

For the purposes of this book, a meadow stream is defined as any stream, no matter the source of its water, with a gentle gradient, meanders, a slow current that primarily takes the form of flats and slick-surfaced runs, stable banks that are usually cut steeply, and at least a minor amount of aquatic vegetation.

The most important water types in meadow streams are flats, smooth runs, and the water along the banks of both, especially at bends. Water types that are often found in meadow streams but are not characteristic include riffles, broken runs, pools, and pocket water. You'll find stretches of these kinds of water along the course of any stream. When you encounter them, they can be important holding water, and you should fish them carefully, according to their kind.

Riffles or broken runs interspersed with more typical meadow stream water can attract trout from long distances. If a short reach of rough water contains the only holding lies safe from overhead predation in a few hundred feet of shallow and placid stream, trout will take up territories there and move out to more dangerous feeding lies only when a hatch is happening. Targeting the broken water of a meadow stream can be a brilliant strategy. Catching trout is somewhat easier in the rougher water, but that's not the main reason it's smart to seek it out. The primary reason is to extend your successful fishing into the many hours when trout are not actively feeding on flat water. The secondary reason is that some of the largest trout on a flat move upstream or drop downstream into safer water types, in a permanent sort of way, as they attain size and need bigger territories.

Flats with smooth surfaces are the kind of water we associate with spring creeks and meadow streams, and it's the kind of water we usually go to these types of streams to fish, because it's only on these sorts of

Meadow streams, especially in their spring creek manifestations, are defined by flats. Their flows are stabilized, the smallest particles—sand and sometimes even silt—are deposited to form the bottom, and vegetation is able to take root and not get swept out annually by the violence of spate flows. They are read by watching for rising trout, visible trout, or the most likely lies—in this case, on Falling Springs in Pennsylvania, marked by the tree that has fallen into the water and offers trout a place to hide from predators.

It's almost always more effective to present your flies to spring creek trout, or any trout on flats, with downstream casts, as Jay Nichols is wisely doing here. Whether you're using dry flies, emergers, or tiny beadhead nymphs suspended beneath small yarn indicators, it's best if your fly arrives in the view of the trout first, ahead of the leader or tip of the line. Quite often, when you cast upstream, the fly lands where you'd like it, two to four feet above the position of the trout you're after, but the leader and part of the line pass over its head and land on the water in its sight. That, in all normal conditions, ends that. Your first order of business when you begin fishing meadow streams, after reading the water, is to learn cross-stream and downstream slack-line casts.

streams that we find this type of water in abundance. It's best to focus on flats here and notice how they meet the basic needs of trout.

The need for shelter from currents is met rather gracefully on meadow stream flats, as the currents are not strong enough to need much breaking.

Plant beds usually give trout all the shelter they need. Shallow trenches worked into the bottom also serve to break the flow. Trenches are found most commonly in meadow streams that flow through volcanic country. Any boulders lodged on the bottom form lies. Logs and limbs are much more likely to lodge in gentle flows of meadow streams than they are in higher-gradient freestone streams. Spring creeks flowing though forests almost always have fallen trees, still attached to the banks by their root systems, lying at angles across their currents. All of these interruptions to the current form lies for trout.

The need for protection from predators is met whenever trout remain on their hidden holding lies, which they choose for the location's ability to provide this as well as shelter from currents. When I make a trip to a typical spring creek, carrying with me my desire to fish over rising trout, I normally make most of my casts to trout that are feeding, few of them to trout holding on protected lies. But that reflects my personality, not yours.

When feeding, trout are almost always on flats or in currents purling along the edges, and their only defense is a superb wariness. They flee on flats at the slightest sign of a bird or fly line passing overhead or a person creeping into their window on the world. The value of a flat as a feeding lie is enhanced considerably by the presence of a nearby bomb shelter, where trout can escape into perfect protection.

On many heavily fished spring creeks, and tailwater flats as well, trout no longer take to their heels at the sight of danger from man. Instead, they go down and turn dour for just a bit, as if they're taking some time to ponder the situation, which they might just be doing. Then they'll slowly regain their composure, rise up and begin feeding once more, right in front of you. Sometimes they won't even take those moments to think things over; they'll just keep feeding on naturals as you approach, take your position, and begin casting to try to catch them. But they know you're there and will be much more difficult to bring to an artificial fly. It's a side note, though an interesting one, that on some such waters we've simply become a part of the local geography to the trout.

Trout foods on meadow stream flats include aquatic and terrestrial insects and small crustaceans. Aquatic insects tend toward a few dominant species, with representatives among the mayflies, caddisflies, and midges most common. Most are small; of these, the midges and blue-winged olives are by far the most prolific. But others are larger, including the pale morning duns, spotted sedges, and sulphurs. Some are very big, such as the western green drakes and brown drakes (*Ephemera*).

Terrestrial insects have importance in proportion to the width of any meadow stream, but it has recently been discovered that all streams of this type have two banks, one on each side, no matter how wide the water, and that the water along those banks is important even if it's a small proportion of the total width of the stream. Because of their relatively constant flows, meadow streams with spring origins tend to be almost as deep at the banks as they are over the rest of their width. Cover is often better at the edge, where grasses and willows overhang the water and offer protection, and the food can be doubled as a result of that meeting of aquatic and terrestrial environments. Wherever a fair current noses along a bank to deliver food, trout will hold there, and terrestrial insects will be a major part of their diet.

Small crustaceans reach their greatest importance in flows with rooted or attached vegetation. Crustaceans love to hike or paddle about in the safety of plants, and they find their groceries there as well. Scuds are most abundant and most interesting to trout in many meadow stream waters. Aquatic sow bugs, also called cress bugs because they thrive in watercress beds, are usually more important on streams with spring sources.

Trout tend to spend their time either feeding visibly on a hatch or other suddenly available bit of food or holding their stations, where they'll usually take whatever food arrives to them on the drift but won't move far for it. When trout feed actively, that is their sole occupation, and it's also when they're most vulnerable to your flies, so long as you can figure out what they're taking, match it at least reasonably, and present your fly as a natural might arrive into their sight. When trout are not feeding, but at rest in holding lies, they probably feed there as willingly as any trout on any station observing the drift.

A holding lie in a spring creek plant bed is a tough place to present a fly so that a resting trout can get a fair look at it. Some plant beds have trenches wide enough to stalk, and to drift a nymph down, without disturbing the trout lying along the bottom. Most of the time it's a tough proposition, and most meadow stream and spring creek anglers prefer to hike and watch, or just wait and watch, for signs of active feeding.

When some sort of insect becomes available in good numbers, trout will move out to take them at whatever level they find them most available. Those trout will be extremely wary when they do, their wariness increasing as they feed nearer and nearer to the surface, exposing themselves to more danger from above.

Tackle and tactics for meadow streams are a quick rehash of those for fishing flats discussed in chapter 8. Rods should be long and light. Lines in the 3- to 5-weight class work best, though joy can be found in even lighter 1- and 2-weights. These are not very useful when the wind comes up, however, and the rods that cast them should be considered specialty rods, not

This is what you're looking for when you read meadow stream flats. It would be beneficial to find a trout rising amid far fewer natural insects than these Trico spinners. This trout was so damnably difficult that I backed away from it, gave up casting, and took photos of it feeding. But I did hook a couple others nearby . . . by *mismatching the hatch*, with a pupa pattern for a size 16 caddis that emerged in sparse numbers among the Tricos. Trout seemed happy to get the occasional caddis natural, and I was happy not to have to watch a size 24 Trico dressing get lost among all those tiny spent spinners.

the ones you reach for when you intend to spend an entire day on a stream and want to be armed for whatever conditions that day might present. Floating lines, weight-forward or double-taper, depending on your desires and what feels best on the rod you choose, are sufficient to fish all meadow stream and spring creek situations. If you need to fish deep, use weighted nymphs, add tiny split shot to the leader, or fix a bit of putty weight around the leader tippet knot. I find myself weighting tiny nymphs with tungsten beads, half the time black and the other half bright, for much of my spring creek fishing.

Leaders should be twelve to fifteen feet long and tapered down to tippets that balance the flies to be cast. Long tippets are a help on smooth water. If you continually change flies, which is a habit most of us have when fishing flats, replace your tippet when it gets shorter than two feet. The difference between an eighteen-inch tippet and a three-foot tippet will be a lot of trout on an average spring creek flat. A longer tippet not only separates the heavier butt of the leader and the tip of the fly line farther from your fly, but also lands on the water relaxed and gives that fly a more natural float. For the same reason, choose soft leader material, rather than stiff, for your spring creek fishing.

You almost need to carry special fly boxes to fish meadow streams and spring creeks. Hatches are specific, and you should be able to match them with reasonable accuracy. It also helps to carry flies for all the levels at which trout feed.

Flies for the bottom should be small nymph dressings, based on the larvae of caddis, the nymphs of mayflies, and imitations for scuds and crustaceans. If caddis are prolific in the waters you fish, carry patterns that match both the larvae and pupae. Stoneflies are not usually abundant in placid flows; they prefer broken water and aren't called stoneflies because they like to live on silty bottoms or in plant beds. Weight on your nymphs will help get them down, but too much weight might sink them too quickly in gentle water. Use beads where they're appropriate, with just a few turns of nonlead wire behind the bead to lock it in. Add weight to the leader if you need to get deeper.

For fishing at mid-depths, lightly weighted nymphs, traditional wet flies, and soft-hackled wets all work well. The best policy is to try to collect a sample of what trout are taking, then select a dressing that resembles it and also has a lifelike action in the water.

Meadow streams and spring creeks with heavy plant beds are rich in a specific challenge: the emerger stage of the aquatic insect. Small mayflies, caddisflies, and even smaller midges bump up against the surface film and often become stuck. It's quite a barrier to penetrate if you're a very tiny insect. Sometimes hundreds of individuals hang just below the surface; they're easy picking for trout. The result is selective feeding that will drive you mad if you try to fish with dry flies that float on the surface rather than flush in it. Nymph dressings with a ball of fur or polypropylene yarn on the back will hang in the surface, just like a natural, and at times can change your luck amazingly. Klinkhamer Special dry flies are so successful because they have no tails to float the back end of the curved-shank hook. It dangles down and looks exactly like a midge pupa hanging from the surface or the nymphal shuck of a mayfly trailing off the body of a newly emerged dun. They're most effective in size 16 to 22, though they work well in larger sizes as well; you just find trout feeding on emergers of smaller insects much more often than larger ones.

When trout do feed on the surface, it will be necessary to fish for them with dry flies that match the naturals they're taking. The smoother the

Meadow streams are defined by their meandering courses. This pastoral stream is the Big Mystery River in Wisconsin, fished here by the most excellent essayist Ted Leeson, author of *The Habit of Rivers*, *Jerusalem Creek*, and *Inventing Montana*. The best holding water is easily read: just follow the currents. The deepest slots are dug in the soft bottom under the most forceful flows. Wherever trout find some slight obstruction to those currents, you'll find prime lies. In this photo, prime lies are located at the point of the little island in the right foreground, where the current makes a turn, at the corner of what is attempting to be a riffle, and all along the undercut bank that Ted is fishing. The bend under Ted's rod forms several more prime lies, which undoubtedly continue down around that bend.

Freestone streams have their occasional meadow stream reaches, which you should approach with the same caution and fish with the same delicacy you would apply to spring creeks of similar size and description. This is the upper Boulder River near Butte, Montana, fished by Trevor Hughes. The river gets its name for the bulk of its course, where it bounds off the nearby Continental Divide. But several miles of its headwaters flow through what must have been rich beaver meadows before those proficient dam builders were trapped out. It's now a meandering meadow stream in that long reach and is best read for lies where obstructions break the current and the current erodes the deepest areas. Trevor hooked this trout where the current kisses the bald, white rock on the right.

water and the more heavily it is fished, the more accurately you'll need to copy the natural. But presentation is at least as important as exact imitation on any smooth water.

On meadow streams and spring creeks, you usually read the trout, not the water, and calculate your presentation based on where they are and what they're doing. First you must determine the stage of the insect they are feeding on and the level at which they are taking it. Then move into position to present your imitation in a way that suits the actions of the natural.

Fishing deep with nymphs calls for upstream presentations. A tiny indicator will help in this kind of fishing. A large, hard one might frighten trout when it lands on the water. On some heavily fished waters, the sight of a chartreuse, yellow, or red indicator floating above a nymph will, based on bad experiences they've had with these things in their past, warn trout about you. Use a white one that looks like foam. You're often better off eliminating the indicator, greasing your leader butt to a point about twice the depth of the water, then concentrating raptly on the floating leader and the end of your fly line. They in effect become your indicator.

Fishing at mid-depths in smooth water usually calls for cross-stream casts, stalking feeding or visible trout so that you can show your fly to them without the leader and line going over them first. Another excellent tactic is to get into position upstream and across from a rising trout. Cast just beyond it and a couple feet upstream from its position. Let the sunk fly swing down and across in front of the trout, and you'll often feel a sullen, satisfying pull. Drop the rod tip if you can muster the discipline, and the hook will set in the corner of the trout's mouth. This is the same hook set used in greased-line Atlantic salmon and summer steelhead fishing, and it takes a lot of practice to get it right. Get it wrong and you often pull the hook away from the fish.

Emerger patterns should be fished across stream, with reach casts, or downstream, with slack-line wiggle casts. These presentations put the fly over the trout ahead of the line and leader, which is the most consistent way to fool trout that are both leader- and angler-shy. Upstream presentations work on rumpled water but not where the surface is smooth.

Dry-fly tactics are the same as emerger tactics, with the reach cast and wiggle cast most effective. It helps in such finesse fishing to insert slack into almost every cast. The slack gives your fly freedom in its drift and defeats microdrag that you can't see. Wobble your rod tip briskly as the line straightens out on the forward cast, and it will land on the water in a series of S turns. As the line straightens on the water, the slack will pull out and the fly will drift without drag.

Whatever kind of fly you use, nymph, wet, emerger, or dry, and no matter the level at which you fish it, bottom, mid-depths, subsurface, or on top, be patient with meadow stream and spring creek trout. It might take many casts before the right fly arrives at the right moment, in the right way, in front of the right trout. When it does, the take will be subtle. Raise the rod slowly and softly to set the hook, then hold on when the trout bores for the plants.

Meadow streams and spring creeks offer a special kind of fishing, with the rewards not necessarily in the size or number of trout you catch, but in the satisfaction of knowing that you beat them at the best of their games.

Spring creeks are some of the most rewarding places to fish in terms of trout caught. Their flats, such as this one on the appropriately named Spring Creek near State College, Pennsylvania, can be read to a certain degree, but I enjoy them most when trout are visible and working, preferably feeding on a hatch of insects or fall of terrestrials. Those are the sorts of situations I go to spring creeks to encounter and possibly solve.

Reading such water consists of spotting the trout more often than it does locating likely lies. Casting to a trout you can see always elevates the level of excitement on such waters. If it's a big one, you need to avoid panic as the trout rises toward your fly.

Tailwater Trout

The attraction of tailwater fisheries for some folks is simple: These waters have an abundance of trout, sometimes including extremely large ones. For other folks, it's more complicated: Many tailwaters have heavy hatches, and those trout, themselves often heavy, rise almost daily and become challengingly selective.

Many tailwater flows, with smaller reregulation reservoirs below power-generating dams to meter out the water, are stable, not subject to winter spate, spring runoff, and summer lows. The temperature of the water, coming as it does from the depths of a reservoir, remains relatively constant. This stability of both flow and water temperature keeps the best tailwaters within the ideal range for insect numbers and trout growth all year round. That is why the trout have potential to grow so large: Food is always available, and they keep eating it all through the year.

Other tailwater flows lack rereg dams and fluctuate daily with power generation needs. The more power that is being generated, the more water flows through the turbines, and the higher the water downstream. When power needs are low, turbines are shut down, and the water downstream from the dam also gets low. The surge as turbines are turned on and the water downstream rises can be fairly abrupt, putting wading anglers in danger. You must keep informed about the sort of tailwater you're wading. If it's one that rises quickly, listen for the whistle that shouts a warning about the rise or watch the water level closely. Fly fishermen drown far too often when they're focused on their fishing and fail to notice the water rising.

Irrigation dams can reverse the seasons on some tailwaters. They store water in late fall, winter, and early spring, then release it during summer crop season. The result is low and fishable flows for the cold half of the year, the season that is normally fished least, and high and often unfishable flows in the season of the year that is normally fished most. Flood control dams restrict normal flows in late winter and spring, taming runoff, and

Tailwaters take many shapes and deliver many sorts of experiences. This is the Marble Canyon stretch of the Colorado River in Arizona, just below Glen Canyon Dam in the background. The trout is Henrietta, named for her similarity in size to the orange cement trout named Henry in Rick Hafele's video *Anatomy of a Trout Stream*. Henrietta took a Gold-Ribbed Hare's Ear nymph fished along the bottom in the slot of wrinkled water that forms where I'm standing, which is about where I hooked her.

The signs that tip you off to good trout lies are often subtle on large tailwaters, but as always, you look for signals on the surface of the way the currents are shaped on the bottom. In this case, the change in the water from smooth all around it to slightly disturbed right in that one spot tells you about boulders on the bottom. These would deflect the minor but constant current and form places for trout such as Henrietta to hold without effort and feed on the bounty that most tailwaters provide.

release it, theoretically at a steady rate, throughout summer and fall. If dam controllers consider fishing important downstream from the dam, they'll stabilize releases to benefit trout—and trout fishermen. If no value is given to the fishing, flows below flood control dams can fluctuate with an odd lack of rhythm that does nothing to help the river and seemingly everything to hurt its fishing.

Every tailwater is different. You'll find it necessary to learn the type of each, and then learn its seasonal and daily rhythms. These vary significantly by region, season, and the nature of power demand. For example, is peak power demand caused by the need for air-conditioning in the afternoon or heat early in the morning? Any literature you can find on a tailwater will help you figure out how to fish it. Many of these waters are so famous they've had books written about them, and angling guidebooks are full of the sorts of information you need. All have had magazine articles extol them, and the Internet offers an abundance of the latest details about every tailwater fishery. Local fly shops, with their business focused on the very tailwater you'd like to fish, have the most recent and reliable information and also follow the hatches daily. You can even find out here which flies to use during different parts of the day, since they have guides out floating the river every day with clients.

The water in a typical tailwater is enriched by planktonic growth in the reservoir above the dam. Plankton does not form in flowing water; it's generally restricted to lakes and ponds, though a very minor amount grows in isolated backwaters and along the rare still edges of streams. Erect a dam on a stream's course, and plankton thrives in the still water formed above it. This richness is delivered to the stream below via the dam's outflow. Populations of aquatic insects that can use plankton as a food source explode to take advantage of the sudden influx of groceries. Trout naturally expand in number and size, taking advantage of the explosive growth of the insects they eat.

Some of the most famous tailwaters are features of arid western country and were warm-water fisheries before the arrival of the dam. The San Juan River in New Mexico, Colorado River in Arizona, and Bighorn River in Montana are examples. But you'll find excellent tailwaters in every region: the White River in Arkansas, the Hiwassee and South Holston in Tennessee, the Bow River in Alberta, on and on all across the continent. You have one near home, and it might be your home water.

My home Deschutes River is a tailwater, with a regulating dam below the dam farthest downstream. Its lower hundred miles have stable flows, but it arises from a giant aquifer beneath the Cascade Mountains and is said to have been more stable, not less so, before any dams were built.

Tailwaters come in almost all sizes, from small trout streams to the largest rivers. Many of the best tailwater trout fisheries arise after a dam is installed on a river at maturity, often where it has spread out, silted in, and warmed up. When such a river is dammed, the reservoir serves as a settling basin for silt, and the depth of the water in the reservoir causes cool flows for some miles downstream.

The result is a river that has some aspects of both freestone and meadow streams. The bottom structure might be that of a freestone stream,

The broad and often brawling Deschutes River, fished here by Robert Sheley in a heavy spring rain that prompted even heavier march brown and blue-winged olive hatches, is a tailwater with four dams along its length. The lowest of them, a hundred miles upstream from the river's confluence with the Columbia, has a reregulating dam. Power is generated, but the resulting unstable flows are stored behind the rereg dam, and a fairly constant flow is released into the lower river. Though changes in levels have been abrupt enough to flood an occasional camp, for the most part you need not worry about rises so swift they put your life in danger. Changes in flow do, however, unsettle the trout, usually putting them off their feed for a day or so.

Stabilized flows on any river, whether from a reregulating dam, spring creek sources, or a lake in the headwaters, reduce spate-type erosion and allow the carving of rather steep banks. You don't find many gravel bars on the Deschutes River or any other with constant flows. That is why this river and many others like it offer excellent bank fishing. Robert is fishing riprap sloughed off a railroad track. The big boulders form constant eddies and other sorts of prime lies.

---

though it remains the type found in a typical lower-river system, with pebble and sand bottoms more common than boulder and stone. If the river had a bouldered bottom before the dam was built, it will have the same after the dam. On many tailwaters that were subject to heavy silting before the dam was built, clean flows leaving the reservoir serve to wash silt downstream, cleansing the riverbed over time.

If the tailwater has stable flows, not subject to scour, rooted vegetation and attached algae both get a chance to take hold. This gives the tailwater aspects of a rich spring creek. This is perhaps the most important feature of the best tailwater fisheries: vegetation that enriches the river with food for trout.

All water types are important on the wide array of tailwater types. Riffles tend to be large, as they are on large rivers. Stability allows the attachment of some kinds of growth, even in fast water. Tailwater riffles can be extremely rich in insect life. If there are places for trout to hold in such riffles, they'll be among the best places to fish. Riffles will rarely hold the largest trout in a tailwater, because the territories are usually too small to satisfy the largest trout. The first sanctuary water downstream from a major tailwater riffle, however, can hold the river's largest trout.

Runs can be very rich in tailwater streams. They offer the protective holding lies typical of freestone streams but have the added benefit of vegetation and more insect life. A freestone run is not often as rich in insect life as a riffle; a tailwater run with rooted vegetation is often much more productive of both insect and trout life than a freestone riffle.

Pools are prime in tailwaters, if the tailwater is shaped so it has pools, though like pools in all big rivers, they can be few and very difficult to fish. If you adjust your equipment to suit the size of the water and are patient enough to do the casting required to wait out a big trout in big water, your chance to take a trophy trout is better in a tailwater pool than it is in any other single water type.

Flats are frequent on tailwater streams. On most tailwaters, they're the definitive water type. Some riffles and runs are tamed to flats by the stable flows below dams. Because of the richness of insect life and consequent heavy hatches, trout rise freely and often on tailwater flats. This is the kind of water we have learned to associate most often with tailwater trout fishing. But there are many tailwaters on which flats as a water type are no more prevalent, and no more important, than flats on typical freestone streams, though it's often the few flats on such tailwaters that get the most fishing pressure. In such situations, some smart anglers seek out runs and pools nearby and find the largest trout holding in them unmolested.

Pocket water is found in at least a few parts of most tailwater fisheries. Where it occurs, it is richer than pocket water found elsewhere, because a tailwater provides increased drift. As a consequence, a pocket is likely to hold larger trout than it would in a typical freestone stream. The same vegetation that causes pocket water to hold larger trout also makes the same

The Hiwassee River tailwater in Tennessee, fished here by Nashville expert Roger Hamiter, has its holding water defined at low flows by base-rock shelves that cut laterally across the river and form trenches parallel to the flows. Boulders also break the current in places, though they're more often protrusions from the bedrock than isolated rocks. Wherever the current is broken and the water downstream is deep enough to provide protection from overhead predation, trout will be found. But the most consistent lies are formed in the broken currents below shelves, such as those where Roger is fishing.

pockets much more difficult and dangerous to reach. Felt soles are not nearly as effective for wading boulders covered with algae as they are in the clean-stoned rapids and cascades of a stream where the algae is washed out each winter. Wear studded brogues and keep that staff in your hand.

The banks of tailwater fisheries, if they're shaped right to meet the three primary needs of trout, are very likely to hold larger trout than the same banks along a stream without a dam. The banks are rarely good unless the tailwater has stable flows. The constant water level stabilizes the erosion pattern and deepens the water next to the banks. Instead of the wide gravel bars of normal freestone streams, which existed before the dam was built, there is more deep water with a constant flow right at the edges. Trout are enticed to hold along the transition line, because the aquatic life is rich there, and the chance of terrestrial life toppling in makes it even richer.

The needs of trout are met in the various water types of a tailwater in exactly the same ways they were described in the earlier chapters on each of those water types. Trout will seek shelter from currents in the same ways and same kinds of places. They will move into feeding lies on a riffle that doesn't have adequate shelter for holding lies, in response to an abundance of food in the riffle. And they'll move back out of the riffle when the food source is gone, seeking shelter once again in holding lies downstream.

The need for protection from predators is also met in the same ways as on all freestone and meadow streams. Trout will be comfortable in deep runs and pools, extremely wary when feeding on exposed flats. They'll choose their holding lies for shelter from currents, but they'll get winnowed out of them if they don't find protection from predators. Water that looks as though it should be full of trout, because it has plenty of feed, can be empty of them until a hatch is going on, when that feed suddenly becomes easier to get. Then the water can boil with feeding trout.

The need for food in a tailwater is met by essentially the same types of insects found in the same water types on freestone or meadow streams. Because of the constancy of conditions in a tailwater, the aquatic insect species list will be narrower, while the abundance of some of those remaining species borders on dominance. That's why it's very important to be aware of the insects and collect samples when you fish tailwater fisheries.

You'll find mayflies, caddisflies, and midges all abundant in tailwater streams. Stoneflies will be present in a few but absent or not important in most. The mix of aquatic insects can change dramatically when a dam is placed on a river. Some species that were present but not thriving before the dam explode a few years after emplacement of a reservoir upstream. Net-spinning caddisfly larvae spin nets like spiderwebs, suspend them in the current to capture tiny bits of drift, and feed on what the net has captured. They exist in scattered numbers in all streams and are sometimes found in good numbers in both freestone and meadow waters. But they expand exponentially in water enriched by plankton, because those tiny bits of food delivered out of the reservoir are just the right size for their nets.

Midge populations also explode in many tailwaters. They find myriads more niches in rooted and attached vegetation, and also in somewhat silty bottoms, than they do in freestone flows. Tame a swift stream with a dam, and midges will take advantage of the more peaceful flows that form downstream. They're usually tiny, size 16 to 24.

Terrestrials are not of any greater importance in tailwaters than they are in freestone streams of similar size. But they are a factor, and trout will feed selectively on them when they are suddenly abundant. You should carry imitations of small ants and beetles, as well as large dressings for grasshoppers.

The Bighorn River below Chief Yellowtail Dam in Montana is one of the continent's most famous tailwaters. It has a reregulating dam that stabilizes its flows. That allows aquatic plants to take root, and these promote an enormous abundance of aquatic insects, mostly mayflies, caddis, and midges, plus crustaceans such as aquatic sow bugs and scuds, and even thriving numbers of aquatic worms. I once hoisted a rock off the bottom of the river. It was encrusted with an inch-thick layer of tightly bound plants that we usually call moss, though they are not. I'm not as educated about aquatic plants as I should be, but these formed what might be termed a crew cut on the rock. I did not see any aquatic insects in the plant material and at first considered it somewhat barren. But after I held it in the air for a few minutes, which makes aquatic life unhappy, things began to move all over it. The rock was crawling with sow bugs, squirming with worms. It was not necessarily the best feeling, holding it, unless you happen to favor the sorts of life that make trout grow large.

Scuds are extremely important in plant and algal beds below dams. Their populations are often intense; they can be the primary food source for trout throughout the winter months, when few aquatic insects except midges hatch and terrestrials are absent at the banks. Aquatic sow bugs take on outsize importance on some Ozark and other tailwaters. On the Bighorn River, I've hoisted rocks covered with what we call moss, but is really attached aquatic vegetation, and found those rocks crawling with sow bugs and aquatic worms.

Tailwaters are often open to fishing through the winter. Winter insects and crustaceans can cause trout to feed constantly and often selectively. With their stable flows and comfortable temperature regimes, tailwaters can offer excellent fishing throughout the year.

Trout spend their time in tailwaters doing the same things they do in other streams. They hold on their stations, feeding on the drift. There are more frequent hatches, so they're more often tempted into moving higher in the water column and off their territories toward feeding lies. They're also more apt to feed all year round, because constant water temperatures keep aquatic insects hatching and trout in their comfort range and active.

Tackle and tactics for tailwaters usually fall into two categories, though all that has been written in this book about fishing the various water types applies exactly to tailwaters, wherever on them you find the particular water type described. The two most important types of fishing on tailwaters are hatch-matching on smooth runs and flats and fishing with big gear on big water and along the banks.

Hatch-matching gear and tactics should be the same as described in chapter 8 for flats and chapter 14 for meadow streams and spring creeks. Rods should be long and light, lines either double-taper or weight-forward floaters in the 3- to 5-weight class. Leaders should be twelve to fifteen feet long, tapered to long tippets that balance the size flies being used.

The flies chosen should match the hatch as nearly as you can and should be appropriate for the level at which you'll fish them. Presentations should suit the insect stage trout are taking and the level at which they're taking it.

Big gear should be the same as that described in chapter 7 for fishing pools and chapter 13 for large rivers. The rod should be long and stout and is used to punch out weight-forward floating lines in 6- to 8-weights. Sinking lines should be wet-heads in various sink rates, or you might want to try the shooting-taper systems so commonly used by those who cast big flies over big water.

The flies should match the biggest bites you expect trout to see. Woolly Buggers and Muddlers and Bunny Leeches are effective, tied with tungsten

Clarks Creek in Pennsylvania is a medium-size tailwater in a forested water-shed. It has many flats, smooth runs, and glides that make it seem almost like a spring creek. But it also has parts that are more like a freestone stream. Sometimes, as in this photo, the two types of water are adjacent. Here a long, glassy flat falls over a couple boulder shelves, forming a short run that is easy to read. Trout hold in the currents just below the shelf, in the deeper water under the leaning trees on the far bank, and might move out onto the tailout of the run to feed if insects emerged there or terrestrials made miscalculated landings on that water. It's a tailwater, but its lies are read just like those on any other stream: for the way the water meets the needs of trout.

beadheads on weighted hooks in size 6 up to a huge 2/0. Keep the leader short if you're going to fish with a sinking line. If you're fishing shallow along shorelines, use a floating line and a leader about the length of the rod.

You'll usually fish this kind of gear in deep tailwater runs, big pools, or when banging the banks. Your tactics for deep water should suit that kind of water, with casts that give the fly lots of time to sink and retrieves that bring it scooting back near the bottom. For banks, give the fly a moment to sink, then begin a short retrieve, lift, and cast again.

The chance of a large trout exists almost anywhere in a tailwater. Because hatches can be so prolific, the largest trout in a given river often set up high in the water column to feed on them. Be prepared to handle a surprise large trout, should one decide to take even your tiniest offering, any time you place a fly on or in a tailwater.

Conclusion

It was our first day on the river. We had thirteen miles of the Bighorn River to float before it got dark, but Jim Schollmeyer and I forgot about time and got hung up casting to rising trout on the flats just downstream from the dam. Trout rose all around us, tilting up slowly out of plant bed lies to sip tiny olive mayfly duns that arose from the same source, then sinking down again, out of sight until the next soft rise.

"It's like the Henrys Fork," I called to Jim as I lifted the fly to cast again after another refusal.

"Except the trout are bigger," Jim called back as he set the hook into a fat trout that mistook his size 20 Olive Compara-Dun for the real thing.

We recalled lessons learned on lots of other rivers, fishing over lots of rising trout. We presented our flies from upstream, with slack-line casts. Often, but never often enough, everything gathered itself into perfection, and a trout came gullibly to an imitation as if it were a natural. We both did well, though Jim did better.

We'd floated only about four miles down the river when Jim suddenly looked up from his intent fishing to another pod of rising trout and said in alarm, "It'll be dark in two hours, and we've still got almost ten miles of river to float!" We jumped into the drift boat and took turns putting our backs to the oars, urging the boat in a losing race against darkness. As we rowed, we watched a lot of river go by that we would rather have been fishing: choppy riffles, long runs, a few deep pools.

"We've got to get off those flats a lot sooner tomorrow," I told Jim, "so we have time to fish some of this other water."

Jim stroked downstream in the dark. "I agree," he answered.

The river turned out to be pretty tame. We laughed as we splashed through the only set of rapids marked on the map. "You sure this is it?" Jim kept asking.

"Must be," I answered. "It's just to the left of that island." I pointed. "And there's the bluff above it." I pointed again.

"Wouldn't be a rapid on the Deschutes," Jim said, and we ceased to worry about going down in the dark.

The next day, we got off the flats on schedule but got hung up in the riffles. They were like the rich, broken water that we fish constantly on our

I have a slide show built around reading trout water. I end it with this slide and tell folks that if they've paid attention, they can catch trout anywhere. The truth is, you can learn a bit about reading trout water from a glance into a stream, even when it's empty of water. In this case, on Big Indian Creek where it comes off Oregon's Steens Mountain, it's easy to see that the prime lie would be in that deep pool just off the grassy point, which would provide protection from overhead predation while deflecting currents outward, forming a bit of a corner, perhaps an eddy, if water were only there and trout were in it . . . which, surprisingly, they are each spring and early summer.

Deschutes River floats nearer to home. We fished them about the same way, with casts quartering upstream and across and size 16 Deer Hair Caddis drys keyed loosely to a natural we noticed flying around over the water. Trout rose and popped spray into the air; we had no trouble telling takes.

This time it was I who cried, "It's almost dark and we've got eight miles to go; let's get!" This time while one of us rowed, the other dapped a dry downstream from the boat in the failing, then later failed, light. We weren't too worried about getting off the river before dark; we knew the trail now.

Dapping downstream was a simple matter of reading seams. Wherever currents came together in that broad river, we put the fly where it would ride right down the suture. Not all seams produced, but each of us took about three more trout than we would have caught had we just rowed.

Since we didn't stop the boat to play the trout, dapping didn't cost us any extra time.

The following day, we parked the boat next to a long run, forced to wait while another party fished a riffle where we'd had good luck the day before and wanted to try again. The run was only about three feet deep and restricted in width by an island, or it would have spread out and been a flat. It had an even but forceful flow for a couple hundred yards. It reminded me of a run on a river I fished a lot at home, and I recalled a time when wet flies had worked wonders for me there when nothing was going on and no hint told about any particular rig or method I should try. While we waited for that other boat to move out of the riffle, I tied on a size 14 Light Cahill wet fly and fished it slowly down the run with wet-fly swings.

The places where trout rapped the fly were almost predictable. There were slight boils on the surface every few feet, marking underwater boulders no larger than basketballs. Whenever the fly swung into position to show itself just over a boulder, I would concentrate, mentally almost coaxing the trout. It worked. Every other boulder or so down the entire length of that run put up a trout. A few of them weighed a couple pounds.

Fishing in that run was so good that when the other boat moved out of the riffle, we neglected to slide right in behind it. We got hung up far above the takeout ramp again as night started to fall and didn't get back until long after dark. On the way down, Jim strung a stout rod while there was light left and tied on a fat and weighted Bunny Leech. He punched it out next to the banks, sometimes bouncing it off rocks and pulling it out of clumps of grass, while I rowed. He caught two trout before it became my time to fish, his turn to take up the oars. Each of his trout weighed more than three pounds.

Jim was a guide before he became a fly-fishing photographer and then a writer. He knows how to position the boat to fish the best water. He also knows how to scold a client into fishing the precise right spots in the best water. As darkness approached, Jim took us through a broad, almost featureless riffle. I was casting at random when Jim suddenly directed, "Cast across that seam." He pointed up toward the bank. "There's a slot in there." I thought he knew about it because he'd fished it before, but he hadn't; he just read it as a place where the water was shallow all around but deepened out briefly.

I cast across the slot to the shallow side of it, mended line a couple times to get the Bunny Leech a bit deeper, then pulled up tight against the streamer, which unfortunately had hooked up on a rock. I made a long roll cast, trying to pull the fly off the rock in the opposite direction. The roll petered out short of the hookup, so I threw a more vigorous roll out there, trying to pull the fly free . . . and you know the rest of the story. I failed to get the fly free and the rock took off. It was a rainbow trout, probably twenty-six inches long then, though it grew quickly after we released it.

The Bighorn is one of the best trout rivers in the world. On that first trip, Jim and I saw anglers from New Jersey and Georgia and Texas and California and everywhere in between, some having excellent fishing, others having fishing no better than they might have had at home. Those who did well read the river carefully, arrowed right to the best water, and fished each water type with a method that suited it. Those who didn't do well wandered down the river, fishing it all equally and with the same method. Even worse, many stopped in one spot and fished it for the rest of the day without ever moving—and, it seemed, without moving many trout.

You can learn what to look for on a river by reading books. They will tell you about the needs of trout and how moving water fills those needs. But the ultimate lessons are learned on trout water. The ultimate instructors are the trout.

Every lesson you learn about reading trout water, no matter where you learn it, transfers to all other creeks, streams, and rivers, no matter where you fish them. Your assignment now: Go fishing.

Bibliography

Bergman, Ray. *Trout*. New York: Alfred A. Knopf, 1938.

Brooks, Charles. *The Trout and the Stream*. New York: Crown Publishers, 1974.

Cutcliffe, H. C. *Trout Fishing on Rapid Streams*. South Molton: W. Tucker, 1863.

Gladwell, Malcolm. *Blink*. New York: Back Bay Books, 2005.

Hafele, Rick. *Anatomy of a Trout Stream*. St. Paul, MN: 3M Scientific Anglers, 1986 (video and companion book).

———. *Aquatic Insects and Their Imitations*. Boulder, CO: Johnson Books, 1987.

———. *Nymph Fishing Rivers and Streams*. Mechanicsburg, PA: Stackpole Books, 2006.

Hafele, Rick, and Dave Hughes. *The Complete Book of Western Hatches*. Portland, OR: Frank Amato Publications, 1981.

Howell, Kevin, and Don Howell. *Tying and Fishing Southern Appalachian Trout Flies*. Pisgah Forest, NC: Davidson River Outfitters, 1999.

Hughes, Dave. *Handbook of Hatches*. 2nd ed. Mechanicsburg, PA: Stackpole Books, 2005.

Hynes, H. B. N. *The Ecology of Running Waters*. Toronto: University of Toronto Press, 1970.

Judy, John. *Slack Line Strategies*. Mechanicsburg, PA: Stackpole Books, 1994.

LaFontaine, Gary. *Caddisflies*. New York: Nick Lyons Books, 1981.

Leeson, Ted, *The Habit of Rivers*. New York: Lyons and Burford, 1989.

———. *Inventing Montana*. New York: Skyhorse, 2009.

———. *Jerusalem Creek*. Guilford, CT: Globe Pequot, 2002.

Nemes, Sylvester. *The Soft-Hackled Fly*. Old Greenwich, CT: Chatham Press, 1975.

Ovington, Ray. *Tactics on Trout*. New York: Alfred A. Knopf, 1969.

Rosborough, E. H. *Tying and Fishing the Fuzzy Nymphs*. 4th ed. Harrisburg, PA: Stackpole Books, 1988.

Schollmeyer, Jim, and Ted Leeson. *The Fly Tier's Benchside Reference*. Portland, OR: Frank Amato Publications, 1998.

Schwiebert, Ernest. *Trout*. New York: E. P. Dutton, 1978.

Swisher, Doug, and Carl Richards. *Fly Fishing Strategy*. New York: Crown, 1971.

Stewart, W. C. *The Practical Angler*. Edinburgh: A. & C. Black, 1857.

Weamer, Paul. *Pocketguide to Pennsylvania Hatches*. New Cumberland, PA: Headwater Books, 2009.

Willers, W. B. *Trout Biology*. Madison: University of Wisconsin Press, 1981.

Index

Page numbers in italics indicate illustrations and sidebars.